Linear Algebra for Data Science with Python

Linear Algebra for Data Science with Python provides an introduction to vectors and matrices within the context of data science. This book starts from the fundamentals of vectors and how vectors are used to model data, builds up to matrices and their operations, and then considers applications of matrices and vectors to data fitting, transforming time-series data into the frequency domain, and dimensionality reduction. This book uses a computational-first approach: the reader will learn how to use Python and the associated data-science libraries to work with and visualize vectors and matrices and their operations, as well as to import data to apply these techniques. Readers learn the basics of performing vector and matrix operations by hand but are also shown how to use several different Python libraries for performing these operations.

Key Features:

- Teaches the most important concepts and techniques for working with multi-dimensional data using vectors and matrices.
- Introduces readers to some of the most important Python libraries for working with data, including NumPy and PyTorch.
- Demonstrates the application of linear algebra in real data and engineering applications.
- Includes many color visualizations to illustrate mathematical operations involving vectors and matrices.
- Provides practice and feedback through a unique set of online, interactive tools on the accompanying website.

John M. Shea, PhD is a Professor in the Department of Electrical and Computer Engineering at the University of Florida, where he has taught classes on stochastic methods, data science, and wireless communications for over 25 years. He earned his PhD in Electrical Engineering from Clemson University in 1998 and later received the Outstanding Young Alumni award from the Clemson College of Engineering and Science. Dr. Shea was co-leader of Team GatorWings, which won the Defense Advanced Research Project Agency's (DARPA's) Spectrum Collaboration Challenge (DARPA's fifth Grand Challenge) in 2019; he received the Lifetime Achievement Award for Technical Achievement from the IEEE Military Communications Conference (MILCOM) and is a two-time winner of the Ellersick Award from the IEEE Communications Society for the Best Paper in the Unclassified Program of MILCOM.

Chapman & Hall/CRC
The Python Series

Image Processing and Acquisition using Python, Second Edition
Ravishankar Chityala and Sridevi Pudipeddi

Python Packages
Tomas Beuzen and Tiffany-Anne Timbers

Statistics and Data Visualisation with Python
Jesús Rogel-Salazar

Introduction to Python for Humanists
William J.B. Mattingly

Python for Scientific Computation and Artificial Intelligence
Stephen Lynch

Learning Professional Python Volume 1: The Basics
Usharani Bhimavarapu and Jude D. Hemanth

Learning Professional Python Volume 2: Advanced
Usharani Bhimavarapu and Jude D. Hemanth

Learning Advanced Python from Open Source Projects
Rongpeng Li

Foundations of Data Science with Python
John Mark Shea

Data Mining with Python: Theory, Applications, and Case Studies
Di Wu

A Simple Introduction to Python
Stephen Lynch

Introduction to Python: with Applications in Optimization, Image and Video Processing, and Machine Learning
David Baez-Lopez and David Alfredo Báez Villegas

Tidy Finance with Python
Christoph Frey, Christoph Scheuch, Stefan Voigt and Patrick Weiss

Introduction to Quantitative Social Science with Python
Weiqi Zhang and Dmitry Zinoviev

Python Programming for Mathematics
Julien Guillod

Geocomputation with Python
Michael Dorman, Anita Graser, Jakub Nowosad and Robin Lovelace

BiteSize Python for Absolute Beginners: With Practice Labs, Real-World Examples, and Generative AI Assistance
Di Wu

Data Clustering with Python: From Theory to Implementation
Guojun Gan

Linear Algebra for Data Science with Python
John M. Shea

For more information about this series, please visit: https://www.routledge.com/Chapman--HallCRC-The-Python-Series/book-series/PYTH

Linear Algebra for Data Science with Python

John M. Shea

CRC Press
Taylor & Francis Group
Boca Raton London New York

CRC Press is an imprint of the
Taylor & Francis Group, an **informa** business

A CHAPMAN & HALL BOOK

First edition published 2026
by CRC Press
2385 NW Executive Center Drive, Suite 320, Boca Raton, FL 33431

and by CRC Press
4 Park Square, Milton Park, Abingdon, Oxon, OX14 4RN

CRC Press is an imprint of Taylor & Francis Group, LLC

© 2026 John M. Shea

ISBN: 978-1-032-65916-9 (hbk)
ISBN 978-1-032-66405-7 (pbk)
ISBN: 978-1-032-66408-8 (ebk)

DOI: 10.1201/9781032664088

Typeset in CMR10 font
by KnowledgeWorks Global Ltd.

Publisher's note: This book has been prepared from camera-ready copy provided by the author.

Dedicated to my parents, Larry and Agnes Shea. Your love and support have shaped who I am and inspire me every day.

Contents

Preface

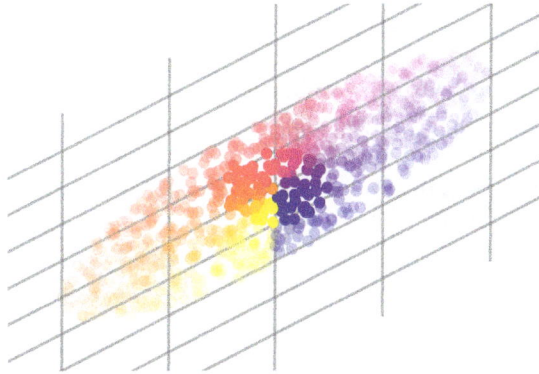

This book is an introduction to linear algebra and its use in data science, including storing and transforming data, solving systems of linear equations, performing data interpolation and regression, and extracting features for dimensionality reduction. This book is targeted toward anyone who wants to learn about linear algebra and its applications, given background knowledge of algebra and basic computer programming. This book fits a unique niche among books on linear algebra:

- **This book applies a modern, computational approach to work with data**.

  - Many books on linear algebra are focused on the mathematical details of matrix and vector manipulation and on proofs of related properties. By using computational tools, this book instead focuses on **how** and **why** to apply the techniques of linear algebra in data science, while also developing these techniques from first principles.

- **This book uses real data sets.**

  - Many linear algebra books use contrived examples that are small enough to print in a book and work with using a calculator, but this results in data sets that are unrealistic and uninteresting. The computational approach used in this book allows the use of real data sets that are too large for manipulation by hand.

- **This book shows how to work with some of the most important tools in the Python data-science stack, including:**

  - NumPy and PyTorch for working with vectors and matrices numerical functions,
  - Pandas for loading, manipulating, and summarizing data, and
  - Matplotlib for plotting data.

- **This book was written alongside the book *Foundations of Data Science with Python*, which covers statistics, probability, and their application to data science using the Python data-science stack.**

  – Techniques like dimensionality reduction combine concepts from probability and statistics, like mean and covariance, with techniques from linear algebra, like eigendecomposition. Dimensionality reduction is covered in each of these texts with minimal coverage of those concepts that are not specific to that particular book.

- **This book provides a unique set of online, interactive tools to help students learn the material, including:**

  – self-assessment quizzes,

  – flashcards to aid in learning terminology,

  – Python widgets and animated plots.

Interactive elements are available on the book's web site: la4ds.net, which can also be accessed using the QR code below:

1

Introduction

Welcome to *Linear Algebra for Data Science with Python*! This chapter introduces the book and its place in the field of linear algebra and data science. It then provides a brief introduction to some of the tools that are used throughout the book. By the end of this book, you will learn:

- the fundamental operations on vectors and matrices,

- how to use matrices and vectors to understand and solve systems of linear equations,

- the interpretation of matrix multiplication as a transformation between vector spaces,

- how a square matrix can have special vectors called *eigenvectors* that preserve their direction when multiplied by that matrix,

- how to use matrix techniques for exact and approximate data fitting, and

- how to use alternate basis vectors to extract useful or important features from data.

While this book does show how to perform many vector and matrix operations and manipulations by hand, the emphasis is on the meaning and application of the techniques. Techniques to perform these operations are always shown in Python using common Python data science libraries.

1.1 Who is this book for?

Although this book is generally designed for scientists and engineers, it should be accessible to anyone who knows:

- algebra and trigonometry,

- some computer programming language (knowing Python is helpful, but not required), and

- complex numbers (minimal knowledge, in Chapter 6 only).

This book is written by an engineer with degrees in both electrical and computer engineering. This book and its companion, *Foundations of Data Science with Python,* were written to provide the main textbooks for a 4-credit, semester-long course for engineers, taught in the Department of Electrical and Computer Engineering at the University of Florida. These books are intended to be a broad introduction to data science, but they are also designed to replace courses in *Engineering Statistics* and *Computational Linear Algebra.*

DOI: 10.1201/9781032664088-1

1.2   Why learn linear algebra from this book?

This book focuses on the most fundamental techniques of linear algebra as they are applied in engineering and the sciences, with a special emphasis on applications to data science. Rather than emphasizing hand manipulation and theory, this book focuses on how to use Python libraries to work with vectors and matrices and how to use these techniques for data science and engineering applications. Using the computer to perform linear algebra not only represents how linear algebra is used in practice but also allows us to work with real data sets that are too large for manipulation by hand to be practical. In addition, the use of computer techniques allows us to better visualize data and the effects of linear algebra techniques.

Interactive flashcards and self-assessment quizzes are provided on the book's website and linked throughout the book to help learners master the material and check their understanding. The entire set of interactive materials can be accessed on the book's website at la4ds.net.

The interactive materials use spaced repetition to help readers retain knowledge as they progress through the book. Starting with Section 2.7, the interactive chapter reviews also give a random subset of review problems from earlier chapters. Research shows that spaced repetition improves the retention of material.

The online materials also include a list of "take-aways" that help summarize the important points from each chapter.

1.3   Brief Introduction to Data Science Terminology

This book focuses on introducing the use of linear algebra techniques for data science. This section reviews some of the basic terminology used when discussing *data*:

> **DEFINITION**
>
> **data**
>
> Collections of measurements, characteristics, or facts about a group.

A simple definition of data science is:

> **DEFINITION**
>
> **data science**
>
> The process of extracting meaning from data.

Data consists of *data points*:

DEFINITION

data point

A collection of one or more pieces of information collected about a single individual or entity.

Each data point may contain *variables* and *features*:

DEFINITIONS

variables

Particular characteristics, measurements, or facts that make up a data point.

features

Individual pieces of information in a data set. While variables typically represent unprocessed or raw data, features can include both variables and processed versions of the variables.

In the machine-learning (ML) literature, the term *feature* is often used for both raw and processed data, especially if the data are used as the input for some ML process.

Variables and features may be either *quantitative* or *qualitative*:

DEFINITIONS

quantitative data

Numeric data. Quantitative data may be either discrete (such as the number of people in a family) or continuous (such as grade point average).

qualitative data

Non-numeric data. Qualitative variables are generally non-numeric categories that data may belong to (such as hair color). Some categories may have an order associated with them, but the order does not imply a numeric nature to the categories. For example, a survey question may have responses from Strongly Disagree to Strongly Agree.

Linear algebra is focused on the application of mathematical techniques to quantitative variables.

Examples of quantitative variables:

- height
- weight
- yearly income
- college GPA
- miles driven commuting to work

- temperature

- wind speed

- population

Readers interested in qualitative variables and statistical tests involving such variables can refer to the book *Foundations of Data Science with Python*, also by John M. Shea.

Terminology review and self-assessment questions

Interactive flashcards to review the terminology introduced in this section and self-assessment questions are available at la4ds.net/1-3, which can also be accessed using this QR code:

1.4   What topics from linear algebra does this book cover?

This book provides an introduction to some of the most important concepts in linear algebra for data science:

- Chapter 2 covers vectors and vector operations, with a special emphasis on correlation and projection. Vector correlation provides a measure of how similar two vectors are, and vector projection is used in alternate representations that extract important information from data.

- Chapter 3 covers the fundamentals of matrices and matrix operations, with emphasis on understanding matrix-vector multiplication as transforming vectors from one vector space to another vector space. The concepts of determinant and eigenvalue-eigenvector pairs are introduced based on this linear transformation viewpoint.

- Chapter 4 examines the application of matrices to solving systems of linear equations. It explores different cases that arise in systems of linear equations, and multiple approaches are presented to solve such systems of equations.

- Chapter 5 covers the common problem of finding the best linear or polynomial fit for a set of data. Techniques are given to find exact polynomial fits for small data sets and approximate polynomial fits for larger data sets. An application to multiple linear regression is shown.

- Chapter 6 introduces the concept of representing data using different bases. Here, a basis is a minimal set of vectors that can be linearly combined to produce every vector in some set. The concepts of universal and set-specific bases are introduced, and an algorithm to find a set-specific basis is given. Three different types of bases are applied to problems in signal processing, digital communications, and dimensionality reduction.

Self-assessment questions

Interactive self-assessment questions are available at la4ds.net/1-4, which can also be accessed using this QR code:

1.5 What topics from linear algebra does this book *not* cover?

This book focuses on introducing some of the most fundamental ideas and operations in linear algebra for use in data science. However, the field of linear algebra is broad, and this book does not cover many important topics. A few important omissions include:

- I do not provide much discussion of vector spaces and especially do not cover the row space or column space of a matrix, nor the associated null spaces.

- Matrix decompositions, in which a matrix is written as a product of other matrices with special properties, are a very important concept in linear algebra. Gilbert Strang, one of the premier educators on linear algebra, considers there to be seven fundamental matrix decompositions. This book only focuses on one: eigendecomposition, which can only be applied to certain square matrices. I briefly touch on three others that are closely related to the material covered in this book: the LU, QR, and SVD decompositions.

- I have omitted most discussion of computational issues, such as the complexity and accuracy of matrix operations carried out using a computer.

- Many data science problems involve sparse matrices, in which the majority of the entries in the matrices are zeros. Because of the importance of sparse matrices in big data problems, there are special ways to efficiently store and operate on sparse matrices.

Self-assessment questions

Interactive self-assessment questions are available at la4ds.net/1-5, which can also be accessed using this QR code:

1.6 Extremely Brief Introduction to Jupyter and Python

The purpose of this section is to briefly introduce users to Jupyter and a few core concepts from Python that will be needed to use Python for linear algebra. The content here should be treated as an introduction to explore further and is not meant to be comprehensive. There are a broad variety of tutorials on the web for both of these topics, and links are provided for users who need additional instruction.

If you are already familiar with Jupyter and/or Python 3, feel free to skip ahead. Similarly, if you are using this as a companion to the book *Foundations of Data Science with Python*, you will find most of the material in this section redundant, with the exception of the information on PyTorch.

1.6.1 Why Jupyter notebooks?

According to the Project Jupyter web page (https://jupyter.org), "The Jupyter Notebook is an open-source web application that allows you to create and share documents that contain live code, equations, visualizations and narrative text. Uses include: data cleaning and transformation, numerical simulation, statistical modeling, data visualization, machine learning, and much more".

The reasons that Jupyter notebook was chosen for this book include:

- Jupyter notebooks can integrate text, mathematics, code, and visualization in a single document, which is very helpful when conveying information about data. In fact, this book and *Foundations of Data Science with Python* were written together in a series of over 140 Juypter notebooks.

- Jupyter notebooks allow for an *evolutionary approach* to code development. Data processing can start as small blocks of code that can then be modified and evolved to create more complex workflows.

- Jupyter notebooks are commonly used in the data science field.

1.6.2   Why Python?

Python is a general-purpose programming language that was originally created by Guido van Rossum and is maintained and developed by the Python Software Foundation. Python was chosen for this book for many reasons:

- **Python is very easy to learn.** Python has a simple syntax that is very similar to C, which many engineers and scientists will be familiar with. It is also easy to transition to Python from MATLAB scripting, which many engineers will be familiar with.

- **Python is an interpreted language,** which means that code can be run directly with immediate feedback and without having to go through extra steps of compiling programs.

- **Python interpreters are freely available and easy to install.** In addition, Python and Jupyter are available on all major operating systems, including Windows, MacOS, and Linux.

- **Python is popular for data science and machine learning.** Python is widely used for data science and machine learning in both industry and universities.

- **Python has rich libraries for linear algebra and data science.** Python has many powerful libraries for data science and machine learning. In addition, Python has powerful libraries for a broad array of tasks beyond the field of data science, which makes learning Python have additional benefits.

1.6.3   How to Get Started with Jupyter and Python

Python and Jupyter are often packaged together in a *software distribution*, which is a collection of related software packages. The creators of several Python software distributions include additional Python software libraries for scientific computing. This book assumes the use of the Anaconda distribution, which its creators bill as "The world's most trusted open ecosystem for sourcing, building, and deploying data science and AI initiatives"[1].

Anaconda's *Individual Edition* is freely available to download from the Anaconda website at https://www.anaconda.com/products/individual. Choose the proper download based on your computer's operating system. You may also have to select a version of Python. This book is based on **Python 3**, and any version of Python that starts with the number 3 and is at least as large as 3.6 should work with the code included in this book. For instance, as of July 2024, the Anaconda distribution included Python version 3.12.

---

[1]https://www.anaconda.com/, retrieved July 19, 2024.

> **WARNING**
>
> ⚠ Python version 2 or Python versions after 3 may have syntax changes
> that cause the code in this book to not run without modification.

After downloading, install Anaconda however you usually install software (for instance, by double-clicking on the downloaded file). Anaconda will install Python and many useful modules for data science, as well as Jupyter notebook and JupyterLab.

> **Note:**
>
> The term "Jupyter notebook" refers to a file format (with *.ipynb* extension), while "Jupyter Notebook" (with a capital N) refers to an application with a web interface to work with those files. To help avoid confusion, I will write *Jupyter notebook file* or simply *notebook* whenever referring to such a file, and we will use JuypterLab as the web application for opening and working with such files.

As of July 2024, JupyterLab "is a next-generation web-based user interface for Project Jupyter" (from https://jupyterlab.readthedocs.io/en/stable/). The Jupyter Notebook application offers a simple interface for working with notebooks and a limited number of other file types. JupyterLab has a more sophisticated interface and can include many different components, such as consoles, terminals, and various editors. The interface for working with notebooks is similar in both, and most users will be able to use them interchangeably.

1.6.4 Getting Organized

We are almost ready to start using Jupyter and Python. Before you do that, I recommend you take a minute to think about how you will organize your files. Learning linear algebra for data science requires actually working with vectors, matrices, and data and performing analyses. This will result in you generating a lot of Jupyter notebook files, as well as some data files. I suggest that you create a folder for this linear algebra book (or for the course if you are using this as a course textbook). This folder should be easily accessible from your home directory because that is the location where JupyterLab will open by default if you use the graphical launcher. You may wish to add additional structure underneath that folder. For instance, you may want to create one folder for each chapter or each project. If you create separate folders for the data, I suggest you make them subfolders of the one containing the notebooks that access that data.

An example layout is shown in Fig. 1.1

1.6.5 Getting Started in Jupyter

Let's begin exploring JupyterLab using an existing notebook:

1) Download a Jupyter notebook file

We will use the file "jupyter-intro.ipynb", which is available at the website for *Foundations of Data Science with Python*:

https://www.fdsp.net/notebooks/jupyter-intro.ipynb

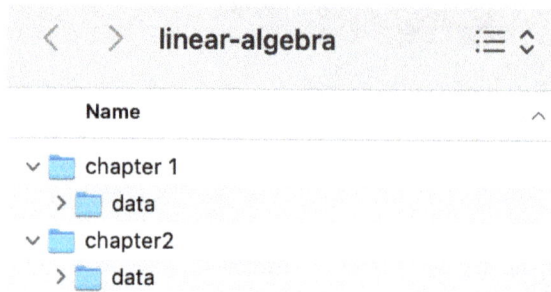

Fig. 1.1: Example directory structure for organizing files for working through the examples and exercises in this book.

If your browser displays the notebook as text, you will need to tell it to save it as a file. You can usually do this by right-clicking or control-clicking in the browser window and choosing to save the page as a file. For instance, in Safari 14, choose the "Save Page As…" menu item. Be sure to name your file with a `.ipynb` ending.

Hint: If your file was saved to your default Downloads folder, be sure to move it to an appropriate folder in your `linear-algebra` folder to keep things organized!

2) Start JupyterLab

JupyterLab can be started from the Anaconda-Navigator program that is installed with the Anaconda distribution. Start Anaconda-Navigator, scroll to find JupyterLab, and then click the *Launch* button under JupyterLab. JupyterLab should start up in your browser.

Alternative for command-line users: From the command prompt, you can start Jupyter-Lab by typing `jupyter lab` (provided the Anaconda bin directory is on the command line search path). Because setting this up is specialized to each operating system and command shell, the details are omitted. However, details of how to set up the path for Anaconda can be found at many sites online.

Your JupyterLab should open to a view that looks something like the one in Fig. 1.2.

WARNING

If you have used JupyterLab before, it may not look like this – it will pick up where you left off!

The JupyterLab interface has many different parts:

1. The **menu bar** is across the very top of the JupyterLab app. I will introduce the use of menus later in this lesson.

2. The **left sidebar** occupies the left side below the menu bar. It includes several different tabs, which you can switch between by clicking the various icons on the very far left of the left sidebar. In Fig. 1.2, the folder icon (■) is highlighted, which indicates that the **File Browser** is selected. For this book, we will use the left sidebar only to access the **File Browser** (■).

3. The **main work area** is to the right of the left sidebar. The main work area will usually show whatever document you are working on. However, if you have not opened

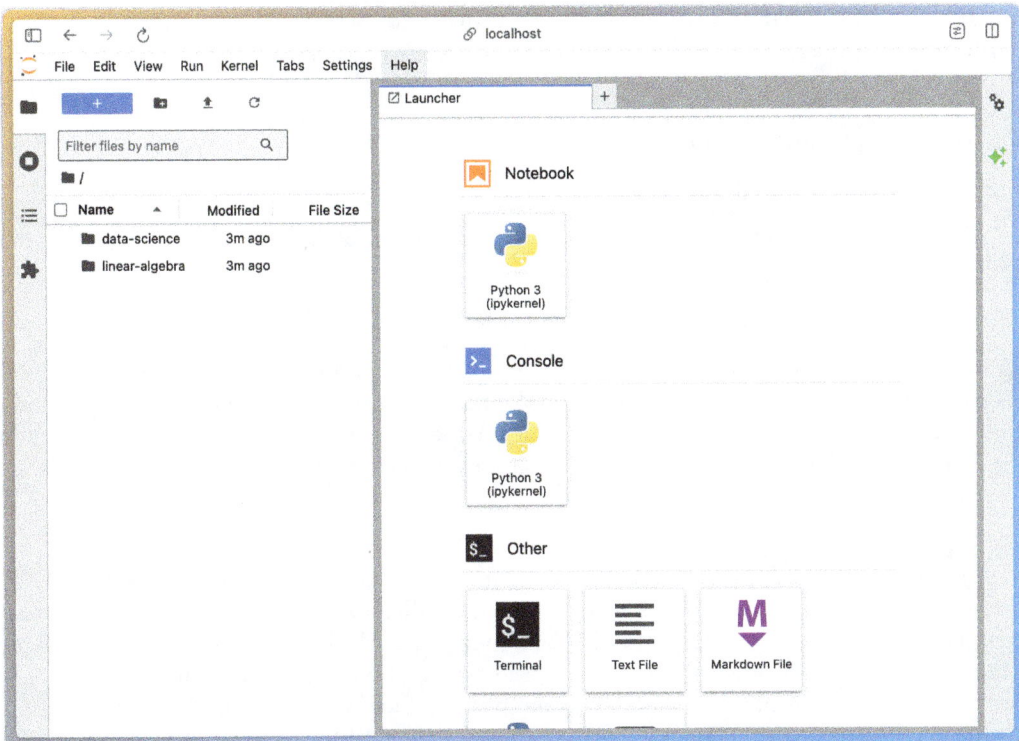

Fig. 1.2: The JupyterLab interface.

any document yet, it will show you different types of notebooks that you can open and other tools that you can access. To start a completely new Jupyter notebook file that can run Python 3 code, you could click on the Python 3 icon under `Notebooks`. For now, you do not need to do that.

Detailed documentation for JupyterLab is available at

https://jupyterlab.readthedocs.io/.

3) Navigate to the downloaded notebook

Use the **File Browser** (■) in the left sidebar of JupyterLab to navigate to the downloaded file.

- *If the **File Browser** (■) is not already showing your files, click on the folder icon (■) on the very left-hand side of the window to switch to it.*

Navigation using the file browser should be similar to navigating in most file selection boxes:

- Single-click on items to select them.

- Double-click on a folder to navigate into it.

- Double-click on a file to open it.

- As you navigate into folders, the current path (relative to your starting path) is shown above the file list. You can navigate back out of a folder by clicking on the parent folder's *name* in the current path.

If you downloaded the file `jupyter-intro.ipynb` to the `chapter1` subdirectory of the `data-science` directory, which lies in your home directory, then you would:

- Double-click on the `linear-algebra` folder.

- Double-click on the `chapter1` folder.

- Double-click on the file `jupyter-intro.ipynb`.

  The file `jupyter-intro.ipynb` should open in the main work area.

1.6.6 Learn the Basics of JupyterLab

After opening the `jupyter-intro.ipynb` notebook, take a minute to scroll through the notebook before interacting with it. Note that the notebook includes formatted text, graphics, mathematics, and Python programming code. Although this book focuses on using Python code for linear algebra, I provide information on the other features because they are useful for documenting and explaining your work.

Notebook structure

Jupyter notebooks are subdivided into parts called **cells**. Each cell can be used for different purposes; we will use them for either Python code or for Markdown. Markdown is a simple markup language that allows the creation of formatted text with math and graphics. Code cells are subdivided into Input and Output parts. Click on any part of the `intro.ipynb` notebook to select a cell. The selected cell will be indicated by a color bar along the entire left side of the cell.

JupyterLab interface modes

The JupyterLab user interface can be in one of two modes, and these modes affect what you can do with a cell:

- In **Edit Mode**, the focus is on one cell, which will be outlined in color (blue on my computer with the default theme), and the cursor will contain a blinking cursor indicating where typed text will appear.

- In **Command Mode**, you cannot edit or enter text into a cell. Instead, you can navigate among cells and use keyboard shortcuts to act on them, including running cells, selecting groups of cells, and copying/cutting/pasting or deleting cells.

There are several ways to switch between modes:

In **Command Mode**, here are two ways to switch to **Edit Mode** and begin editing a cell:

- Double-click on a cell.

- Select a cell using the cursor keys and then press $\boxed{\text{Enter}}$.

  In **Edit Mode**, here are two ways to switch to **Command Mode**:

- Press $\boxed{\text{Esc}}$. The current cell is *not evaluated*, but it will be selected in **Command Mode**.

- If editing a cell that is *not* the last cell in the notebook, press $\boxed{\text{Shift}}+\boxed{\text{Enter}}$ to *evaluate* the current cell and return to **Command Mode**. (If you are in the last cell of the notebook, $\boxed{\text{Shift}}+\boxed{\text{Enter}}$ will evaluate the current cell, create a new cell below it, and remain in **Edit Mode** in the newly created cell.)

More on cells

In **Edit Mode**, code or Markdown can be typed into a cell. Remember that each cell has a *cell type* associated with it. The cell type does not limit what can be entered into a cell. The cell type **determines how a cell is evaluated**. When a cell is evaluated, the contents are parsed by either a Markdown renderer (for a Markdown cell) or the Python kernel (for a Code cell). A *kernel* is a process that can run code that has been entered in the notebook. JupyterLab supports different kernels, but we will only use a Python kernel. Cells may be evaluated in many different ways. Here are a few of the typical ways that we will use:

- Most commonly, we will evaluate the current cell by pressing $\boxed{\text{Shift}}$ + $\boxed{\text{Enter}}$ or $\boxed{\text{Shift}}$ + $\boxed{\text{Return}}$ on the keyboard. This will always evaluate the current cell. If this is the last cell in the notebook, it will also insert a new cell below the current cell, making it easy to continue building the notebook.

- It is also possible to evaluate a cell using the toolbar at the top of the notebook. Use the triangular "play" button (pointed to by the red arrow in the image in Fig. 1.3) to execute the currently selected cell or cells.

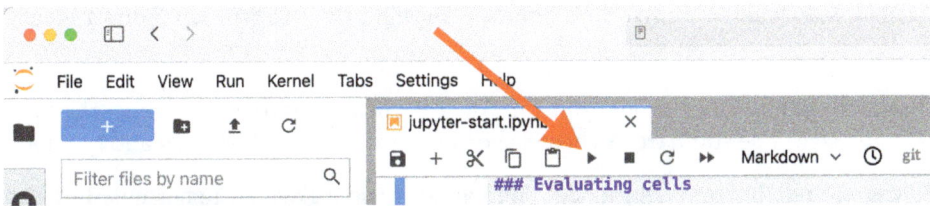

Fig. 1.3: Image of Jupyter interface indicating location of "play" button for executing cells.

- Sometimes we wish to make changes in the middle of an existing notebook. To evaluate the current cell and insert a new cell below it, press $\boxed{\text{Alt}}$ + $\boxed{\text{Enter}}$ on the computer keyboard.

- Cells can also be run by some of the commands in the `Kernel` menu in the JupyterLab menu. For example, it is always best to reset the Python kernel and run all the cells in a notebook from top to bottom before sharing a Jupyter notebook with someone else (for example, before submitting an assignment). To do this, click on the `Kernel` menu and choose the `Restart Kernel and Run All Cells...` menu item.

If you enter Markdown into a Code cell or Python into a Markdown cell, the results will not be what you intend. For instance, most Markdown is not valid Python, and so if Markdown is entered into a Code cell, a syntax error will be displayed when the cell is evaluated. Fortunately, you can change the cell type afterward to make it evaluate properly.

> **Important!**
>
> ! New cells, including the starting cell of a new notebook, start as *Code* cells.

Cells start as *Code* cells, but we often want to enter Markdown instead. We also may wish to switch a *Markdown* cell back to a *Code* cell. There are three easy ways to change the cell type:

- As seen in Fig. 1.4, you can use the drop-down menu at the top of the notebook to set the cell type to *Code*, *Markdown*, or *Raw*.

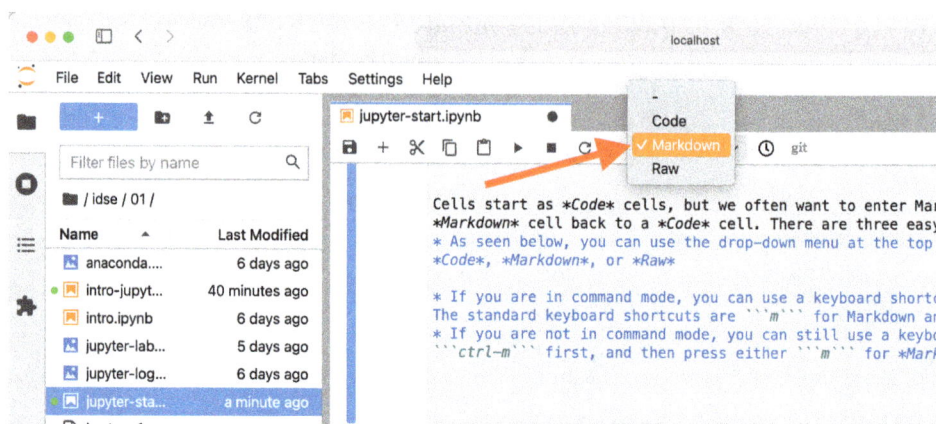

Fig. 1.4: Picture of JupyterLab interface showing the cell type drop-down menu.

- If you are in command mode, you can use a keyboard shortcut to change the type of a selected cell. The standard keyboard shortcuts are ⎡m⎤ for *Markdown* and ⎡y⎤ for *Code*.

- If you are not in command mode, you can still use a keyboard shortcut, but you will need to press ⎡Control⎤+⎡m⎤ first, and then press either ⎡m⎤ for *Markdown* or ⎡y⎤ for *Code*.

Intro to Markdown in Jupyter

This book primarily focuses on explaining linear-algebra techniques and their application to analyzing data using Python libraries. However, it is helpful to not only do the mathematical and data manipulation but also to document your results. Markdown can be used to add text, heading, mathematics, and other graphics.

The example notebook `jupyter-intro.ipynb` demonstrates the main features of Markdown. Recall that you can double-click on any cell in the notebook to see the Markdown source. The `jupyter-intro.ipynb` notebook illustrates the features of Markdown listed below. A tutorial on how to create each of these in Markdown is given online at fdsp.net/1-6.

1. Headings are written like `# Heading`, where more # can be added for subheadings.

2. Text and paragraphs. Paragraphs are indicated by blank lines.

3. Emphasis can be added to text using asterisks, with single asterisks indicating `*italics*` and double asterisks indicating `**bold**`.

4. Bulleted lists can be created by putting items after an asterisk followed by a space: `* my list item`.

5. Numbered lists can be created by putting items after a number, a period, and a space: `1. my numbered item`.

6. Links can be created by putting the link text in square brackets, followed by the link URL in parentheses, like `[Example link](http://google.com)`

7. Images are created in a similar way to URLs, except have an exclamation point (!) before the square brackets: `![Image example]{my_image.jpg}`.

8. Math can be entered using LaTeX notation.

A good reference for Markdown syntax is Markdown Guide: https://www.markdownguide.org/extended-syntax/.

Getting Notebooks into and out of JupyterLab

There are several ways to get notebooks into JupyterLab:

- As previously mentioned, you can use the **File Browser** (■) to navigate to the current location of a file. Note that you will be constrained to only navigating to files in the directory in which Jupyter was started or in any subdirectory below that. One disadvantage of this approach is that your work will be saved wherever that file currently resides. For instance, if you have downloaded a notebook from the internet into your `Downloads` folder, your work on that notebook will remain in the `Downloads` folder.

- You can use drag-and-drop to copy any file into a directory that you are currently browsing using JupyterLab's **File Browser** (■). To do this:

  - Open the **File Browser** (■) in Jupyter and navigate to the directory where you want to work.

  - In your operating system's file manager (e.g., Windows Explorer or Mac Finder), open the folder containing the file you want to copy.

  - Position and resize the folder and your web browser's window so you can see both simultaneously.

  - Click and hold on the Jupyter notebook file that you want to move. Then drag it onto the file list area of the **File Browser** (■).

  - When the Jupyter notebook is over JupyterLab's **File Browser** (■) panel, the outline of the **File Browser** panel will change to indicate that it is ready for you to drop the file. Release the mouse button or trackpad to copy the file into the selected directory.

  - **Note that this makes a copy of the file from its original location.**

- As an alternative to drag-and-drop, you can click on the upload icon (an arrow with a line under it) at the very top of the **File Browser** (■) panel. This will bring up a file selector that you can use to copy a file from anywhere on your computer.

You can save your work by choosing `Save Notebook` in JupyterLab's `File` menu or by pressing the keyboard sequence listed next to that item in the menu. When you manually save your work in this way, Jupyter actually saves two copies of your work: it updates the `.ipynb` file that you see in the file list, and it also updates a *hidden* checkpoint file. When you are editing or running your notebook file, Jupyter will also autosave your work periodically – the default is every 120s. When Jupyter autosaves, it only updates the `.ipynb` file. If Jupyter crashes or you quit it without saving your notebooks, your last autosaved work will be what you see in the `.ipynb` files. However, you can always revert to the version you purposefully saved by using the `Revert Notebook to Checkpoint` item in the `File` menu.

When starting new Jupyter notebooks, their initial name will be "Untitled.ipynb". You can easily rename your notebook in a couple of ways. First, you can choose the `Rename Notebook...` option from the file menu. As an alternative, you can right-click on the notebook in the left-hand **File Browser** (■) panel and choose `Rename`. In both cases, be sure to change only the part of the notebook name that is in front of the `.ipynb` extension. Jupyter uses that file extension to recognize Jupyter notebook files.

Important!

When you are finished working with a Jupyter notebook, I recommend you perform the following steps:

1. First, from the `Kernel` menu, choose `Restart Kernel and Run All Cells...` This will clear the previous output from your work and rerun every cell from the top down.

2. Check over your notebook carefully to make sure you have not introduced any errors or produced any unexpected results from having executed cells out of order or from deleting cells or their contents. By performing these first two steps, you help make sure that someone else loading your notebook file will be able to reproduce your work.

3. Check the notebook file name and update it if necessary.

4. Save the notebook.

5. Choose `Close and Shutdown Notebook` from Jupyter's `File` menu.

6. If you are finished working in JupyterLab, then choose `Shut Down` from JupyterLab's `File` menu.

Another common workflow in JupyterLab is to use an existing notebook as a starting point for a new notebook. Again, there are several ways to do this:

- If you already have the existing notebook open, then you can save it under a new name by choosing `Save Notebook As...` from Jupyter's `File` menu and giving the notebook a new name. **Note that after you use this option, the notebook that is open in the main work area will be the notebook with the new name. You will no longer be working on the original notebook.**

- You can also duplicate a notebook by right-clicking on the notebook's name in the **File Browser** (■) panel on the left-hand side and choosing `Duplicate`. A copy of the notebook will be created with the name of the existing notebook appended with a suffix like `-Copy1` before the `.ipynb`.

Jupyter magics

Code cells can also contain special instructions intended for JupyterLab itself, rather than the Python kernel. These are called *magics*, and a brief introduction to Jupyter magics is available at the website for *Foundations of Data Science with Python* at fdsp.net/1-6.

1.6.7   Getting Started in Python

In this section, I want to introduce a few Python concepts that will be used throughout the following chapters. A more general introduction to Python is available online at fdsp.net/1-6. For users who want to learn more about Python, the following resources are recommended:

- *A Whirlwind Tour of Python* (https://jakevdp.github.io/WhirlwindTourOfPython/) by Jake VanderPlas is a free eBook that covers all the major syntax and features of Python.

- Learn Python for Free (https://scrimba.com/learn/python) is a free 5-hour online introduction to Python (signup required).

- The Python documentation includes a Python Tutorial: https://docs.python.org/3/tutorial/.

Python is an interpreted language, which means that when any Code cell in a Jupyter notebook is evaluated, the Python code will be executed. Any output or error messages will appear in a new output portion of the cell that will appear just after the input portion of the cell (that contains the Python code). At the bottom of the `jupyter-intro.ipynb` notebook, there is an empty cell where you can start entering Python code. If there is not already an empty cell there, click on the last cell and press `Alt-Enter`.

First, Python variables are dynamically typed, meaning that you do not have to specify what type of data they contain. Python variable names must start with a letter and consists of alpha-numeric characters and underscores (_). You can create a Python variable by assigning to it:

```
x = 10
```

If we just want to see the value of a variable, we can evaluate a Jupyter cell containing the variable's name:

```
x
```

```
10
```

More generally, many Python statements return results, and if the last command in a cell returns results, these will appear in a special *output* part of the cell.

We often want to combine some fixed text and some variable output. To do this, we will use Python's `print()` command with *f*-strings, which were added to Python in version 3.6. An *f*-string is a special string that is created by prefixing the first string-delimiter with the letter `f`. Any part of an *f*-string contained within curly braces {} will be evaluated before the string is used. For instance,

```
print(f'The square of x is {x**2}')
```

```
The square of x is 100
```

Lists, Tuples, and Zero-based Indexing

We will often be creating Python representations of two types of linear algebra objects, vectors and matrices, that can be stored as indexed sets of numbers. We will often use a Python `list` object to pass the numbers that make up a vector or matrix to whichever class that we are using to represent that vector or matrix. A `list` is an ordered, mutable store of information; *mutable* means that the contents of a `list` can be changed. A Python `list` is indicated by enclosing the members in square brackets [ ], with items separated by commas. For example, the following code creates a `list` with four elements and then evaluates that `list`:

```
c = [1, 2, 4, 8]
c
```

```
[1, 2, 4, 8]
```

Elements in a list can be retrieved by passing the element index in square brackets after the variable name. Computer languages can usually be partitioned based on how they index the first element in a variable that has multiple elements. Python uses zero-based indexing, which means that the first element in an object with n elements has index 0, and the last element has index $n - 1$. The following code prints the first element (at index 0) and the last element (at index 3):

```
print(c[0], c[3])
```

```
1 8
```

Ranges of elements can be selected using `[a:b]` notation, where `a` is the first element to be selected and `b-1` is the last element to be selected (so that `[0:n]` selects every element in a length n object. For example, if we use the range selector `[1:3]` on c, we get:

```
print(c[1:3])
```

```
[2, 4]
```

More details and examples involving indexing are given in Section 2.1.

The Python `tuple` data type is closely related to the `list` type. Like lists, tuples are ordered collections of data, and indexing works in the same way. Unlike lists, tuples are *immutable*, which means that their contents cannot be changed after creation. A `tuple` is indicated in Python by enclosing its members in parentheses ( ). If there is only one element in a tuple, that element must be followed by a comma to distinguish that the parentheses are used to indicate a tuple rather than to indicate mathematical or logical grouping. In this book, tuples are primarily used for passing collections of parameters to certain functions. Although lists could be used for this purpose, tuples are more common, and so I have chosen to stick with that convention. Here is a tuple version of the list c:

```
ctup = (1, 2, 4, 8)
ctup
```

```
(1, 2, 4, 8)
```

The following shows what happens when we use a range of the form [3:], which means from index 3 to the end. Since there are no members beyond index 3, a tuple with only one member is returned:

```
ctup[3:]
```

```
(8,)
```

Numerical Operations with NumPy

Almost all of our work on linear algebra will utilize the NumPy (usually pronounced "Numb Pie") module, which contains many numerical functions. To use a module, you must import it into your Python working environment. We will use the standard convention of importing the NumPy module into the `np` namespace as follows:

```
import numpy as np
```

To call a function from a namespace, we type the name of the namespace, followed by a period, followed by the name of the thing you are trying to access. For instance, the value of π is a constant object named `pi` in NumPy. Now that we have imported NumPy, we can access that value:

```
print(np.pi)
```

```
3.141592653589793
```

NumPy has many typical mathematical functions, which we can call using the `np` namespace:

```
np.sin(np.pi / 4)
```

```
0.7071067811865475
```

The NumPy class that we will use to represent vectors and matrices is the `ndarray`, or simply NumPy array. We can create a NumPy array from our list `c` by passing it as the sole argument to the function `np.array()`:

```
cn = np.array(c)
cn
```

```
array([1, 2, 4, 8])
```

NumPy arrays also use zero-based indexing, and we can retrieve elements from a NumPy array using the same type of square-bracket notation as for lists:

```
cn[1:3]
```

```
array([2, 4])
```

We will use NumPy arrays instead of Python lists because NumPy arrays have operators that work like vector and matrix operators, and NumPy offers many functions for working with matrices.

Linear Algebra Operations in PyTorch

PyTorch is an alternative to NumPy for most of the work considered in this book. PyTorch is a popular library for machine learning (ML), and many ML algorithms build on linear algebra techniques. Unlike NumPy, PyTorch is not installed by default in Anaconda. PyTorch can be installed from the QT Console (available from the Anaconda Navigator app) or from a terminal as:

```
conda install pytorch::pytorch
```

From within JupyterLab, you can install PyTorch by running a code cell with the command above prepended with an exclamation mark:

```
!conda install pytorch::pytorch
```

The installed library is called `torch` when importing, and the equivalent to NumPy's `ndarray` is PyTorch's `tensor` object. We can create a Python `tensor` from our list `c` as follows:

```
import torch

ct = torch.tensor(c)
ct
```

```
tensor([1, 2, 4, 8])
```

PyTorch tensor indexing is zero-based and works the same as for NumPy arrays:

```
ct[1:3]
```

```
tensor([2, 4])
```

Choosing Between NumPy Arrays and PyTorch Tensors

Both NumPy arrays and PyTorch tensors can represent vectors and matrices, and they share many similar operations. However, they are often used for different purposes and have different advantages:

- NumPy arrays are designed for efficient numerical computing and data analysis. The key advantage of NumPy arrays is their simplicity and wide support in the data science ecosystem.

- PyTorch tensors are specifically designed for machine learning, particularly deep learning. The main advantage of PyTorch tensors is their support of advanced features used in machine learning, such as the ability to run operations on GPUs.

> **Important!**
>
> For the linear algebra concepts covered in this book, both NumPy arrays and PyTorch tensors work equally well. I primarily present examples using NumPy arrays because they are simpler and more widely used in basic data science. However, I also include information on carrying out the same operations using PyTorch tensors, and more details are included on the book's website at la4ds.net.

Converting Between Arrays and Tensors

- To convert a NumPy array to a PyTorch tensor: `torch.tensor(numpy_array)`

- To convert a PyTorch tensor to a NumPy array: `torch_tensor.numpy()`

Objects and Methods

Python variables are more powerful than variables in many languages because they are actually *objects*. Python is an object-oriented programming language. This book does not generally use an object-oriented approach; however, you will need to know some fundamentals about **objects** and **classes**:

- *Objects* are special data types that have associated *properties* and *methods* to work on those objects. Properties are values that are associated with an object. Properties of an object are accessed by giving the variable/object name, adding a period, and then specifying the property name. Methods are similar to functions, except they are specialized to the *object* to which they belong. Methods are called by giving the variable/object name, adding a period, specifying the method name, and then adding parentheses, with any arguments provided in parentheses.

- A *class* is a template for an object that defines an object's properties and methods.

For example, NumPy arrays and vectors have a variety of methods. We can call the `sum()` method of the NumPy vector `cn` that we created previously:

```
cn.sum()
```

15

Loading and Analyzing Data in Pandas

Because this book focuses on the application of linear algebra to data science, we will have the need to load, display, and perform some basic operations on data sets. Although we can load data files directly into NumPy arrays, I instead show how to load data through Python's Pandas library. Pandas is widely used for storing and manipulating small datasets, and Pandas is used for this in the companion book, *Foundations of Data Science with Python*. We will import the Pandas library into the `pd` namespace:

```
import pandas as pd
```

Pandas is for working with tabular data, i.e., data that can be tabulated into rows and columns. We will generally store such data in a Pandas *dataframe*. A dataframe is similar to a spreadsheet or a database table. It is a two-dimensional structure in which each row corresponds to a single data point. Each column stores one variable or feature. Like tables in spreadsheets or databases, the columns can be labeled, and the rows can be indexed by consecutive integers or by the data in one of the columns.

Pandas has a `read_csv()` function for reading comma-separated values (CSV) files or tab-separated values (TSV) files, it has a `read_excel()` function for reading Microsoft Excel files, and it has many other functions for reading data from other statistical software, such as SAS and SPSS, along with functions to read many other file types. The following code reads a CSV file and stores its contents as a Pandas Dataframe object called `brfss`. Then the first five rows are displayed by using the `head()` method of the Dataframe class.

```
brfss = pd.read_csv('https://www.fdsp.net/data/brfss21-hw.csv')
brfss.head()
```

	HTIN4	WEIGHT2
0	59.0	72.0
1	65.0	170.0
2	64.0	195.0
3	71.0	206.0
4	75.0	195.0

We can select data from a particular column by putting column name in square brackets after the variable name for the dataframe:

```
brfss['HTIN4'].head()
```

	HTIN4
0	59.0
1	65.0
2	64.0
3	71.0
4	75.0

When accessing one column of a Pandas `dataframe`, the result is a Pandas `Series` object. A Pandas `Series` is similar to a Python `list`, but each item has its own index value, and the `Series` has many of the same methods as a `dataframe`. We will introduce additional Pandas methods and functions as we need them later in the book.

Self-assessment questions

Interactive self-assessment questions are available at la4ds.net/1-6, which can also be accessed using this QR code:

1.7 Chapter Summary

This chapter introduced the topics that will be covered in this book, as well as JuypterLab and Python, which are two of the main tools used throughout the book. JupyterLab is used to provide a computational notebook environment. These notebooks can combine programming code, text, graphics, and mathematics. We will use these to perform numerical operations, analyze data, and present results. Python is used because it is widely adopted by the data science and machine learning communities, as well as being a general-purpose programming language. Python has well-developed libraries for data science and many other applications, and we introduced a few of the libraries that we will use throughout this book.

Access a list of key take-aways for this chapter, along with interactive flash-cards and quizzes at la4ds.net/1-7, which can also be accessed using this QR code:

2

Vectors and Vector Operation

Vectors provide a way to collect and operate on multiple pieces of numerical data. In this chapter, I define vectors and introduce different ways to visualize vector data. Then I introduce the most common vector operations and their properties. The chapter ends with a discussion of vector projection, which introduces concepts of how we can approximate a vector as a scaled version of another vector. This builds the foundation for our work on transforming data in Chapter 6. Throughout, the concepts are demonstrated through examples using Python and the NumPy library.

2.1 Introduction to Vectors

Let's begin by providing a simple definition of a *vector*:

> **DEFINITION**
>
> **vector**
>
> > An ordered collection of numbers that has an accompanying set of mathematical operations.

Vectors are often used to represent quantities in two-dimensional or three-dimensional Euclidean (regular geometric) space, in which case we can consider vectors to have both magnitude and direction. However, this book takes a much broader view of vectors. Vectors are used throughout science and engineering as a way to store and operate on collections of numerical phenomena. From a data science perspective, we will use vectors to store data points or data features.

We can consider vectors to be a collection of numbers indexed along a single axis. In Chapter 3, we will consider similar mathematical objects (matrices and tensors) for which the collections are indexed across multiple axes. We will call the number of indices the *order*:

> **DEFINITION**
>
> **order**
>
> > The number of axes used to index the contents of a mathematical object, such as a vector, matrix, or tensor.

DOI: 10.1201/9781032664088-2

Thus, all vectors are of order one. For tensors, the order is also sometimes called the rank (although rank has another meaning that is covered in Section 4.2) or degree.

To distinguish between vectors and single numerical values, we call the latter *scalars*:

DEFINITION

scalar

    A single numerical value.

Scalars are considered to be of order zero. As previously mentioned, vectors that represent quantities in Euclidean space can be considered to have both magnitude and direction, whereas a scalar has only a magnitude and sign. We will generally be dealing with scalars and vector components that come from the real line, which we denote by the symbol $\mathbb{R}$.

Notation

!

In this text, vectors are written as bold, lowercase letters, such as $\mathbf{u}$, whereas scalars are written as non-bold, lowercase letters, such as c. In handwriting, vectors should be written as lowercase letters that are underlined.

In other books, vectors are sometimes indicated using a one-barbed, right-pointing arrow over the letter, like $\vec{u}$, especially for geometric vectors that represent a magnitude and direction in Euclidean space.

Vectors are usually represented mathematically as a column of numbers enclosed in large square brackets, like

$$\mathbf{u} = \begin{bmatrix} 0.75 \\ -1 \\ 1.75 \\ 2.5 \end{bmatrix}. \tag{2.1}$$

To save space, we can also write a column vector like $\mathbf{u} = [0.75, -1, 1.75, 2.5]^\mathsf{T}$, where the superscript $^\mathsf{T}$ indicates that the vector should be "transposed" from a row to a column.

Example 2.1: Vector in NumPy

We can create a Python object that represents $\mathbf{u}$ using NumPy's `array` class. For convenience of discussion, we will refer to either the mathematical object or it's Python representation as a vector. To create a Python vector u, we can pass a Python `list` of numbers to `np.array()`:

```
import numpy as np

u = np.array([ 0.75, -1, 1.75, 2.5])
u
```

```
array([ 0.75, -1.  ,  1.75,  2.5 ])
```

In NumPy, the number of dimensions of the array corresponds to the order of the mathematical object that the array represents. We can get the order of u using the `ndim` property:

```
u.ndim
```

1

Example 2.2: Vector in PyTorch

PyTorch's `tensor` class is very similar to NumPy's `array` class in terms of both methods and operators. Thus, we can create a `tensor` object to represent **u** by either passing a list to `torch.tensor` or by passing it a NumPy `array` object. Thus, we can create a PyTorch `tensor` to represent **u** as follows:

```
import torch

u2 = torch.tensor(u)
u2
```

```
tensor([ 0.7500, -1.0000,  1.7500,  2.5000], dtype=torch.float64)
```

As with NumPy, the order of the PyTorch tensor can be retrieved using the `ndim` property of the `tensor` class:

```
u2.ndim
```

1

> **Note:**
>
> For most of the mathematical operations considered in this book, NumPy arrays and PyTorch tensors have the same methods and operators. Thus, I will only present the NumPy version. The PyTorch code is included on the book's website.

A vector consists of *components* or *elements*:

DEFINITION

component (vector),
element (vector)

   One of the numerical values that make up the vector.

We will later generalize vectors to allow the components to be variables that represent numbers.

For a vector $\mathbf{u}$, we will denote its ith component by u_i. Here, i is called the *index* of the component. Consistent with the use of zero-based indexing in Python, we will take the index of the first element as 0. However, the reader should be aware that in general math applications, the first index is often taken as 1.

Example 2.3: Accessing a Component of a Vector in NumPy

For the vector $\mathbf{u} = [0.75, -1, 1.75, 2.5]^\mathsf{T}$ from Example 2.1, $u_2 = 1.75$.

   In NumPy, particular components of arrays can be retrieved using *indexing*, in which an index or set of indices is specified in square brackets following a vector variable. For instance, we can retrieve component 2 of the NumPy vector u as shown in the following:

```
u[2]
```

```
1.75
```

Example 2.4: Accessing a Component of a Vector in PyTorch

   Retrieving a particular component of a vector is slightly different in PyTorch because indexing into a PyTorch tensor always returns a tensor. Consider the following example:

```
u2[2]
```

```
tensor(1.7500, dtype=torch.float64)
```

If we want to retrieve the value of that single-item tensor, we can use the `item()` method:

```
u2[2].item()
```

```
1.75
```

Note that the `item()` method of a PyTorch tensor only works on tensors with one element.

Example 2.5: NumPy Indexing by List

If the index is a list or vector of values, the result will be a vector. For instance, we can retrieve elements 1 and 2 of u as follows:

```
u[[1,2]]
```

```
array([-1.  ,  1.75])
```

Example 2.6: NumPy Indexing by Range

We can also specify a range of consecutive indices as `a:b`, but remember that in Python, the upper end of the range is not included in that range. Thus, to get elements 1 and 2 using range notation, we need to do the following:

```
u[1:3]
```

```
array([-1.  ,  1.75])
```

Another useful way to index vectors by range is to specify a `step`. The `step` is provided as the third component of a range (after a second colon). Note that if no value is provided for the first two components of a range, then the range is assumed to go over all values of the vector. Thus, we can get every second element of u, starting from element 0, as follows:

```
u[::2]
```

```
array([0.75, 1.75])
```

It is often helpful to find out how many elements a vector contains:

> **DEFINITION**
>
> **size (vector),**
> **dimension (vector)**
>
> The number of components a vector contains.

> **WARNING**
>
> ⚠ The size, or dimension, of a vector is also called the *length* in some books, but this may lead to confusion because it is generally not equal to the length of the vector in Euclidean space.

Example 2.7: Size of a Vector in NumPy

The dimension of the vector **u** from Examples 2.1–2.4 is 4.

We will determine the size of a NumPy vector using the `size` property of a NumPy vector: `u.size` or NumPy's `np.size()` function: `np.size(u)`.

```
np.size(u)
```

```
4
```

More generally, we can find the size of any NumPy array across each of its indices using the `shape` property:

```
u.shape
```

```
(4,)
```

Example 2.8: Size of a Vector in PyTorch

PyTorch does not a have a `torch.size()` function, but we can determine the size of a PyTorch tensor using that tensor's `size()` method or `shape` property:

```
u2.size()
```

```
torch.Size([4])
```

```
u2.shape
```

```
torch.Size([4])
```

> Both of these techniques return the same result. The result is a `Torch.Size` object, but it can be treated the same as a tuple for our purposes.

A vector of size n is called an n-dimensional vector, or simply an n-vector. The set of all n-vectors whose components can be any real number (i.e., each $x_i \in \mathbb{R}$) is denoted by $\mathbb{R}^n$.

Terminology review and self-assessment questions

Interactive flashcards to review the terminology introduced in this section and self-assessment questions are available at la4ds.net/2-1, which can also be accessed using this QR code:

2.2   Visualizing Vectors

Vectors are often visualized as displacements from a point, meaning an indication of movement from some starting point to an ending point. If no starting point is given, then the displacement is measured from the origin.

We are only going to plot 2-vectors in this book. For 2-vectors, the components of the vectors are interpreted as representing x and y displacements:

$$\mathbf{a} = \begin{bmatrix} a_x \\ a_y \end{bmatrix}.$$

One of the most common ways to illustrate 2-vectors is to draw each 2-vector as an arrow from the origin to the coordinates given in the vector. This is a special case of a *quiver plot*:

DEFINITION

quiver plot

A (two-dimensional) plot that illustrates one or more vectors as arrows that are typically specified by a location, which determines the coordinates of the tail of the vector, and some specification of the direction and magnitude of the vector. If no location is provided, then the origin $(0, 0)$ is used.

In some applications, *quiver plot* is used to refer to plots where the vectors' locations are at points in a grid, and the vectors indicate some magnitude and direction associated with that location. For instance, such plots are used to illustrate fluid flows over surfaces. I will use the term *quiver plot* to refer to any plot that illustrates vectors as arrows.

The PlotVec Library

!

To make it easier to create plots of vectors as arrows, I have created a library of functions for plotting vectors called PlotVec and made it available via the Python Package Index (PyPI), which is a standard repository for distributing Python libraries across distributions. To install Python packages via PyPI, you can use the `pip` command, which is a text-based (not graphical) command. In most cases, you can run `pip` commands from within JupyterLab by prefixing them with an exclamation point. To install the PlotVec library from within JupyterLab, you can run the following command in any code cell:

```
!pip install plotvec
```

If `plotvec` is already installed, you can upgrade to the latest version as follows:

```
!pip install -U plotvec
```

If you have trouble, you can review additional information on installing libraries via pip at https://packaging.python.org/en/latest/tutorials/installing-packages/.

In all future sections that rely on the `plotvec()` function, I will assume that the `plotvec` module has been installed.

The PlotVec library contains two functions for plotting vectors, `plotvec()` and `plotvecR()`. I will include parentheses after the function names to help distinguish them from the library name. Both functions have essentially the same purpose: plot one or more vectors as arrows. By default, the `plotvec()` function enforces an equal aspect ratio on the vector plot, which means that a unit of length will occupy the same amount of visual space in both the horizontal and vertical directions. This is often useful when trying to illustrate the relation among multiple vectors; however, we do not need to enforce equal aspect ratios for the plots in this section. We could pass the keyword argument `square_aspect_ratio=False` to `plotvec()`; however, a more concise alternative is to use the function `plotvecR()`, which uses a rectangular aspect ratio by default.

Assuming that you have installed the PlotVec library, import the `plotvecR()` function into your global namespace:

```
from plotvec import plotvecR
```

Example 2.9: Plotting a Vector with plotvecR()

Consider the vector $\mathbf{a} = [2, 3]^\mathsf{T}$. We can visualize this vector using the following interpretation. Since no initial location is specified, start at the origin. Then move 2 in the x direction (to the right) and 3 in the y direction (up). We draw an arrow from (0,0) to (2,3) to represent this Vector:

```
a = np.array([2, 3])
plotvecR(a)
```

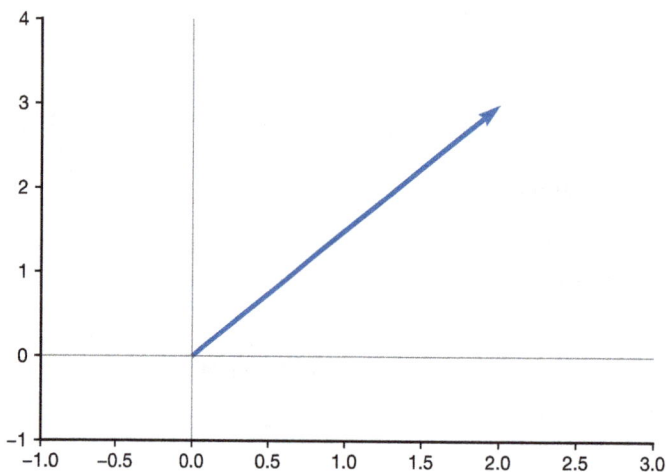

Using this type of visualization, the vector has a direction and magnitude, as previously mentioned. The direction can be measured as the angle of the vector measured from the orientation of the positive x-axis, and the magnitude can be measured as the length of the vector. When plotted as an arrow, the vector is considered to have a *tail* and *head*:

DEFINITION

tail (vector)

> The *tail* of a vector is the starting point of a vector (the initial point from which the displacement is measured).

DEFINITION

head (vector)

> The *head* of a vector is the ending point of a vector (the point at which the vector terminates after the specified displacement from the tail).

For our example, the tail is at the origin $(0, 0)$. Thus, the displacement of $(2, 3)$ results in the coordinates of the head of the vector being the same as the displacement: $(2, 3)$. Thus, we draw the vector as an arrow from the tail at $(0, 0)$ to the head at $(2, 3)$, with the tip of the arrow at the head.

Example 2.10: Plotting Two Vectors with Tails at the Origin

Let's create a second vector. The `plotvecR()` function can handle plotting multiple vectors:

```
b = np.array([1, -2])
plotvecR(a, b)
```

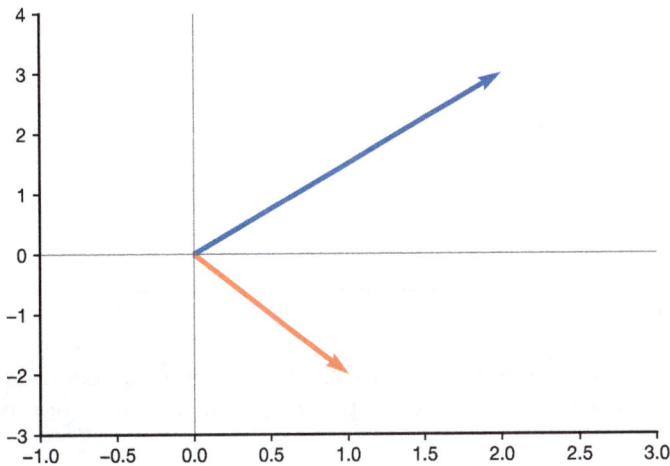

Exercise: Create another vector **c**, and plot it with **a** and **b**. Vary the coordinates of **c** and see how that changes the figure.

Terminology review and self-assessment questions

Interactive flashcards to review the terminology introduced in this section and self-assessment questions are available at la4ds.net/2-2, which can also be accessed using this QR code:

2.3 Applications

Vectors are used in many different ways. Below are some of the ways that vectors are used, along with an example of each type of use:

Geometrical features, such as location in space or displacements between locations

Most of the target audience of this book will already be familiar with this use from physics classes. Vectors can be used to represent physical features of dynamical systems, such as position, velocity, and acceleration in two- or three-dimensional Euclidean space.

In Fig. 2.1, the vectors shown as arrows with solid lines represent the movement of a robot in a plane during two consecutive periods. The robot starts at the origin, which is the tail of the blue vector. It travels to the head of the blue vector during period 1. Its starting position in period 2 is the same as the ending position in period 1, so the tail of vector 2 is located at the head of vector 1. The head of vector 2 is the position of the robot at the end of period 2. The arrow shown with a dashed line indicates the vector from the robot's initial position to its final position; later we will show that this vector is the sum of the two movement vectors.

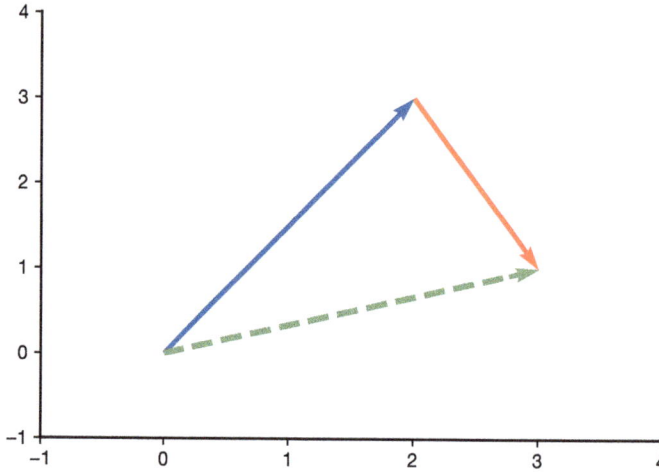

Fig. 2.1: Two solid arrows represent vectors indicating movement of a robot over two consecutive periods. Dashed arrow represents vector pointing to final position of robot.

Multi-dimensional numerical data

We often represent a data set as a table. For instance, each row may represent one data point, and each column may represent one feature. If all of the data features are numeric or encoded as numerical values, then we can use vectors to represent such data in different ways. For instance, each data point can be represented as a vector, or the observed values for a particular feature can be represented as a vector.

Example 2.11: Vector Representations and Plots of Heights and Weights

"The Behavioral Risk Factor Surveillance System (BRFSS) is the nation's premier system of health-related telephone surveys that collect state data about U.S. residents regarding their health-related risk behaviors, chronic health conditions, and use of preventive services": https://www.cdc.gov/brfss/index.html.

The BRFSS 2021 survey contains over 400,000 records and over 300 variables. It takes a long time to load and work with the full set of survey results, so I have extracted data for two variables to analyze. The variables are as follows, and I performed data cleaning for each variable as described:

- **HTIN4**: A computed variable that lists height in inches. Invalid responses ("Don't know/Not sure", "Refused", or "Not asked or Missing") have been dropped.

- **WEIGHT2**: The reported weight in pounds. Again, I have dropped invalid responses, as above.

Even after cleaning, the resulting data set has over 391,000 data points.

We can load the cleaned height and weight data into a Pandas dataframe called `brfss`, as we did in Section 1.6.7. Here, I again show how to load the CSV data into a dataframe and display the first five rows:

```python
import pandas as pd

brfss = pd.read_csv('https://www.fdsp.net/data/brfss21-hw.csv')
brfss.head()
```

	HTIN4	WEIGHT2
0	59.0	72.0
1	65.0	170.0
2	64.0	195.0
3	71.0	206.0
4	75.0	195.0

Because the data occupies a two-dimensional table, we can decompose it into vectors in two different ways: we can treat each row (i.e., data point) as a vector or each column (i.e., feature) as a vector. Let's start by treating each row as a 2-vector and generate a plot of these vectors. Because it would be hard to see and interpret over 391,000 arrows representing all of the rows, let's plot the arrow representations for the first 50 rows. In the following code, I use a `for` loop to iterate over the first 50 rows and call `plotvec()` for each one:

```python
from plotvec import plotvec

for i in range(50):
  plotvec(brfss.iloc[i], color_offset=2*i, square_aspect_ratio=False,
          newfig=False)
plt.xlim(0,80)
plt.ylim(0,350)
plt.xlabel('Height (in)')
plt.ylabel('Weight (lbs)');
```

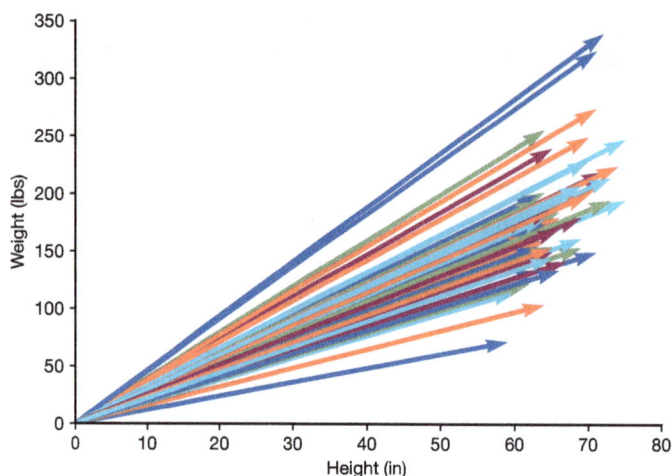

Do you notice any trend in the arrows? We expect them all to point up and to the right because both height and weight are positive quantities. However, you should notice an additional trend that most of the arrows point in the same general direction. This might indicate that these two features are not independent of each other. Intuitively, you may guess that, as a general trend, taller people are more likely to be heavier than shorter people.

If we want to visualize more of the data, then plotting the vectors as arrows is probably not the best approach. When plotting data with two numerical features, it is much more common to plot each data point as a single point in the plane, where one of the features acts as the x-coordinate and the other feature acts as the y-coordinate. Equivalently, we can think that each point is located at the head of the corresponding vector in the arrow plot. Such a plot is called a *scatter plot*:

DEFINITION

scatter plot

A (two-dimensional) scatter plot takes a sequence of two-dimensional data points (x_0, y_0), (x_1, y_1), ..., (x_{n-1}, y_{n-1}) and plots symbols (called *markers*) that represent the locations of the points in a rectangular region of a plane.

With the exception of plotting vectors, we will use the Matplotlib library to create plots. In fact, even PlotVec uses Matplotlib to create quiver plots. The most common way to make plots in Matplotlib is to use the PyPlot submodule, which provides many plotting commands that are similar to those in MATLAB. It is usually imported as `plt`:

```
import matplotlib.pyplot as plt
```

The function `plt.scatter()` can be used to create two-dimensional scatter plots. Instead of accepting a sequence of n two-dimensional data points, which might result in a large number of inputs, `plt.scatter()` expects one n-vector of x-coordinates and one n-vector of y-coordinates.

Thus, an alternative way to represent our two-dimensional height and weight data is to represent each feature by one vector. We can extract all of the height data as a Pandas Series object by passing the column name `'HTIN4'` as the index: `brfss['HTIN4']`. If we need this data as a NumPy vector, we can convert it using the Pandas Series `.to_numpy()` method, like `brfss['HTIN4'].to_numpy()`. However, for the purposes of passing this data to `plt.scatter()`, that is not necessary because `plt.scatter()` can directly accept the Pandas Series.

Even when using a scatter plot, plotting every data point makes a plot that is very large in size when saved as a pdf. Instead, we downsample to every 100th point, using the range notation `::100`. We apply this to both the `'HTIN4'` column as the independent (x-axis) data and the `'WEIGHT2'` column as the dependent (y-axis) data. The following code generates the scatter plot:

```
plt.scatter(brfss['HTIN4'][::100], brfss['WEIGHT2'][::100], 4, alpha=0.7)
plt.xlabel('Height (in)')
plt.ylabel('Weight (lbs)')
```

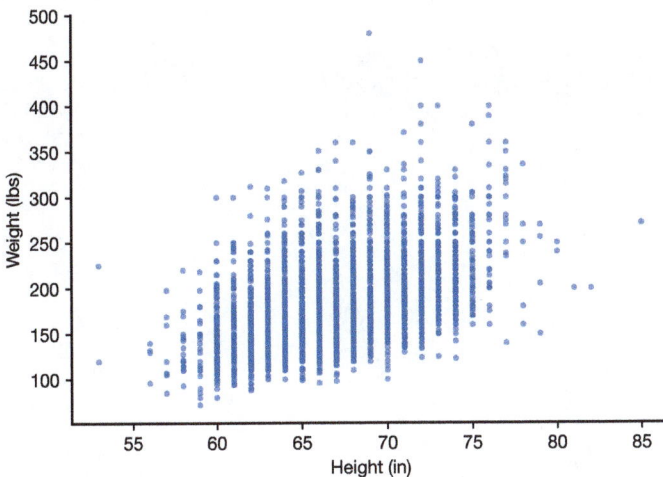

Note that I have used the functions `plt.xlabel()` and `plt.ylabel()` to add appropriate x-axis and y-axis labels, respectively.

The scatter plot shows a similar trend to what we saw in the previous quiver plot. The general trend is that larger heights are generally associated with larger weights.

Time-series data

Many data sets consists of observations of some phenomena over time, and these are called *time-series data*:

> **DEFINITION**
>
> **time-series data**
>
> Data that is collected over time, usually at regular intervals. Each data point is associated with a timestamp indicating when the data was collected.

For instance, the daily closing price of a stock over the past year can be represented by a vector. Some other examples of time-series data include weather and climate data.

Is the climate changing? That is a complicated question that is outside of the scope of this book, but we can try to answer it on a small scale. Since I live in Florida, let's consider the research question, "Is the annual temperature changing over time in Miami-Dade County, Florida?"

Example 2.12: Annual Temperature Data for Miami-Dade County

Let's start by loading the annual temperature data for Miami-Dade County from the National Oceanic and Atmospheric Administration:

```
# Alternate site for accessing data:
# df = pd.read_csv('https://www.fdsp.net/data/miami-weather.csv', skiprows=4)
df=pd.read_csv('https://www.ncei.noaa.gov/access/monitoring/climate-at-a-glance/'
             + 'county/time-series/FL-086/tavg/ann/5/'
             + '1895-2022.csv?base_prd=true&begbaseyear=1895&endbaseyear=2022',
             skiprows=4)
df.head()
```

	Date	Value	Anomaly
0	189512	73.6	-1.1
1	189612	73.9	-0.8
2	189712	74.6	-0.1
3	189812	74.4	-0.3
4	189912	74.7	-0.0

The `Value` column contains the annual temperature. The `Date` column contains the year followed by a two-digit month code, which can be ignored because this is annual data. Let's create a separate column for the year:

```
df['Year'] = df['Date'] // 100
df.head()
```

	Date	Value	Anomaly	Year
0	189512	73.6	-1.1	1895
1	189612	73.9	-0.8	1896
2	189712	74.6	-0.1	1897
3	189812	74.4	-0.3	1898
4	189912	74.7	-0.0	1899

The following code generates a scatter plot showing the annual temperature (on the y-axis) as a function of the year (on the x-axis):

```python
plt.scatter(df['Year'], df['Value'], 15)
plt.xlabel('Year')
plt.ylabel('Annual temperature ($^\circ$F)')
```

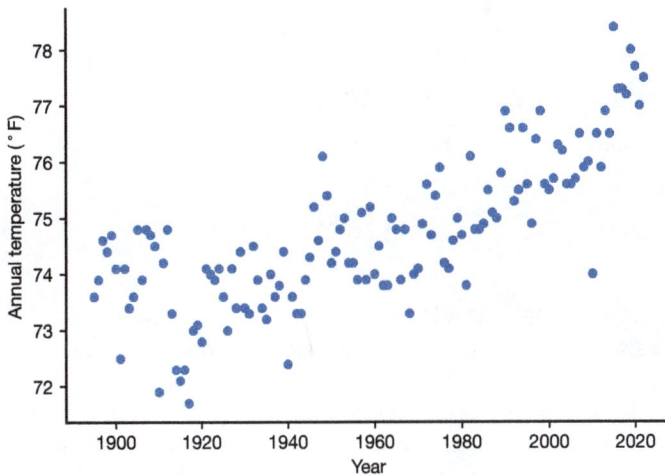

The results seem to show an even stronger relation between annual temperature and year than the relationship between height and weight seen in the BRFSS data.

Distributional data

Vectors may be used to indicate distributions or allocations across categories. For example, an investor's current net worth across different categories (such as stocks, bonds, and real estate) can be represented as a vector. In some cases, we are interested in the proportional allocations, which we can get by dividing the un-normalized distributions by the sum of the entries in the vector.

> ## Example 2.13: Composition of the Dow Jones Industrial Average
>
> The Dow Jones Industrial Average (DJIA) is a stock market index that is based on the prices of 30 major companies traded on the New York Stock Exchange or NASDAQ exchange. This index is computed by adding the trading prices of the 30 companies and then dividing by a factor that is used to compensate for stock splits. The following code loads the price and weight information for the 30 DJIA stocks as of November 26, 2024. (This data was retrieved from SlickCharts Dow Jones Companies: https://www.slickcharts.com/dowjones):

```
dow = pd.read_csv('https://www.fdsp.net/data/dowjones-112624.csv')
dow.head()
```

	Company	Symbol	Weight	Price
0	UnitedHealth Group Incorporated	UNH	8.315803	606.79
1	Goldman Sachs Group Inc.	GS	8.277370	605.50
2	Home Depot Inc.	HD	5.884052	429.52
3	Microsoft Corp	MSFT	5.748436	427.99
4	Caterpillar Inc	CAT	5.568073	407.83

> Here, the `Weight` is the proportion of the index that each stock represents, expressed as a percentage. Stocks with higher prices make up more of the index. Let's extract the `Weight` column as a vector (unlike the previous examples, where we extracted the rows as vectors). Although not necessary for plotting the weights, it is more obvious that the weights can be considered as a vector of proportional data if we extract them into a NumPy vector. The best way to do this is to use the `to_numpy()` method of the Pandas `dataframe` and `Series`, as shown in the following:

```
weights = dow['Weight'].to_numpy()
np.round(weights, 1)
```

```
array([8.3, 8.3, 5.9, 5.7, 5.6, 5.5, 4.7, 4.3, 4.2, 4.1, 4. , 3.6, 3.4,
       3.2, 3.2, 3.1, 2.8, 2.4, 2.2, 2.1, 2.1, 1.9, 1.8, 1.6, 1.4, 1.2,
       1.1, 0.9, 0.8, 0.6])
```

> We can immediately observe that the index is heavily weighted toward the top few stocks. For instance, if we sum the top five weights, we get:

```
np.sum(weights[:5])
```

```
33.793734
```

Thus, the five stocks with the highest prices represent over a third of the index.

When plotting proportions, it is most common to use either bar charts or pie charts. However, there have been many issues identified with pie charts – see, for example, "Why you shouldn't use pie charts": https://scc.ms.unimelb.edu.au/resources/data-visualisation-and-exploration/no_pie-charts. In particular, people are not good at estimating proportions from angles. Therefore, I will use a *bar chart* to illustrate proportional data:

DEFINITION

bar chart,
bar graph

Most commonly used with categorical data for which each category has an associated quantity or measurement, a bar is drawn for each category, where the height or width of the bar is proportional to the associated quantity or measurement.

The following code generates a bar chart that shows the proportion of the index that each of the top 5 stocks represents:

```
plt.bar(dow[:5]['Symbol'], dow[:5]['Weight'])
plt.xlabel('Stock Symbols')
plt.ylabel('Proportion (%)')
plt.title('Proportion of Dow Jones Industrial Average; Top 5 Stocks');
```

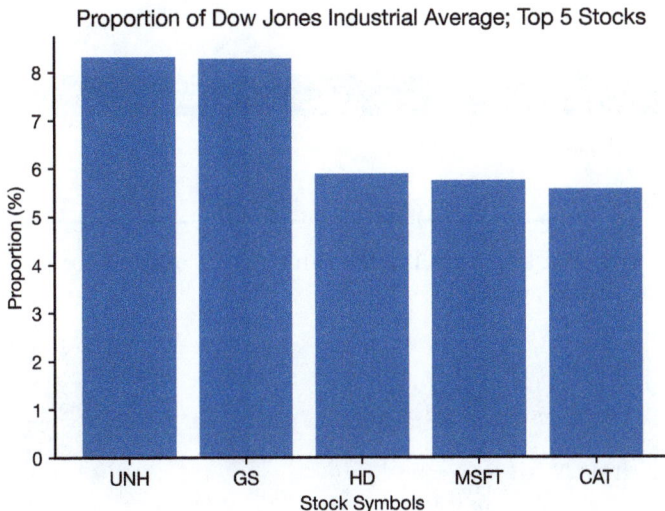

Terminology review and self-assessment questions

Interactive flashcards to review the terminology introduced in this section and self-assessment questions are available at la4ds.net/2-3, which can also be accessed using this QR code:

2.4   Special Vectors

Below are several types of vectors that are common enough to have their own names and notations.

DEFINITION

zero vector

A vector of all zeros.

We will denote the zero vector of size n by $\mathbf{0}_n$. For example,

$$\mathbf{0}_5 = \begin{bmatrix} 0 \\ 0 \\ 0 \\ 0 \\ 0 \end{bmatrix}.$$

We can create a zero vector in NumPy by passing the desired size to `np.zeros()`. Thus we can create a Numpy representation of $\mathbf{0}_5$ as

```
zeros5 = np.zeros(5)
print(zeros5)
```

```
[0. 0. 0. 0. 0.]
```

DEFINITION

ones vector

A vector of all ones.

We will denote the ones vector of size n by $\mathbf{1}_n$. We can create a ones vector in NumPy by passing the desired vector size to `np.ones()`:

```
np.ones(5)
```

```
array([1., 1., 1., 1., 1.])
```

> **DEFINITION**
>
> **standard unit vector**
>
> A vector with all of its components equal to zero, except one element that is equal to one.

For a given size n, we denote the standard unit vector with element i equal to 1 by $\mathbf{e}_i$.
For example, the three standard unit vectors of dimension 3 are:

$$\mathbf{e}_0 = \begin{bmatrix} 1 \\ 0 \\ 0 \end{bmatrix}, \quad \mathbf{e}_1 = \begin{bmatrix} 0 \\ 1 \\ 0 \end{bmatrix}, \quad \text{and} \quad \mathbf{e}_2 = \begin{bmatrix} 0 \\ 0 \\ 1 \end{bmatrix}. \tag{2.2}$$

In NumPy, we can create a standard unit vector by creating a zeros vector of the same size and then setting one desired component to 1. For instance, we can create a NumPy representation of $\mathbf{e}_2$ as

```
e2 = np.zeros(3)
e2[2]=1
print(e2)
```

```
[0. 0. 1.]
```

Terminology review and self-assessment questions

Interactive flashcards to review the terminology introduced in this section and self-assessment questions are available at la4ds.net/2-4, which can also be accessed using this QR code:

2.5   Vector Operations

One of the main reasons to use vectors is that they enable simple notation for, and implementation of, a variety of mathematical operations for collections of numerical data. In this section, we consider these operations and their visualization.

Let's start by loading the `plotvec()` function and visualizing two vectors,

$$\mathbf{a} = [2, \quad 3]^\mathsf{T}, \text{ and}$$
$$\mathbf{b} = [1, -2]^\mathsf{T}.$$

```
import numpy as np
from plotvec import plotvec

a = np.array([2, 3])
b = np.array([1, -2])
```

(continues on next page)

(continued from previous page)

```
plotvec(a, b,
        labels = ['$\mathbf{a} = [ 2,3]^T$',
                  '$\mathbf{b} = [ 1, -2]^T$'],
        legendloc='upper left', square_aspect_ratio=False)
```

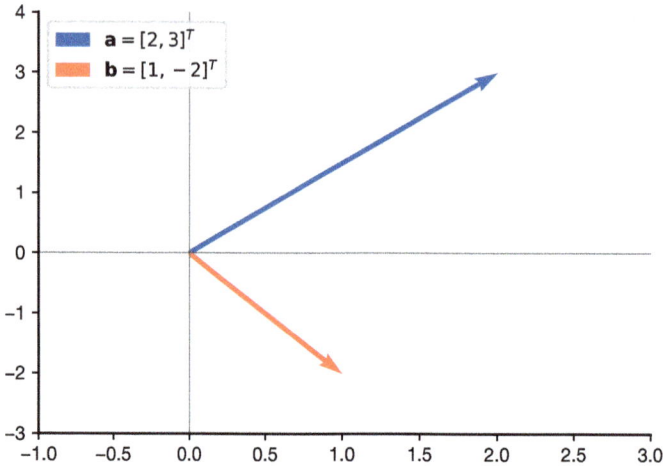

2.5.1   Vector Addition

Vector addition is one of the easiest operations because it is just component-wise addition:

DEFINITION

addition (vectors)

The sum of n vectors $\mathbf{a}$ and $\mathbf{b}$ is a vector $\mathbf{a} + \mathbf{b}$ such that any component i of the vector sum is the sum of the ith components of $\mathbf{a}$ and $\mathbf{b}$; i.e.

$$(\mathbf{a} + \mathbf{b})_i = a_i + b_i, \quad i = 0, 1, \ldots, n - 1.$$

We can visualize the sum of a series of vectors as the displacement achieved from the consecutive (chained) displacement of the vectors, where the head of each vector serves as the tail of the next vector in the series. The `plotvecR()` function will chain vectors in this way if given the keyword argument `chain=True`:

```
plotvecR(a, b, chain=True)
```

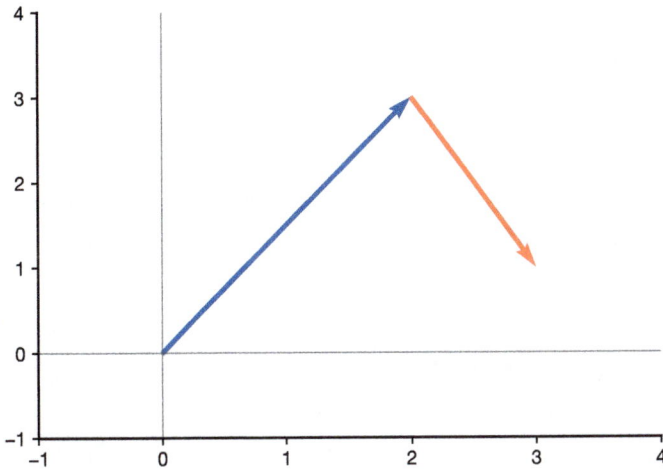

The `plotvecR()` function can also plot the vector from the original tail to the final head if we specify the `plotsum=True` parameter:

```
plotvecR(a, b, chain=True,  plotsum=True)
```

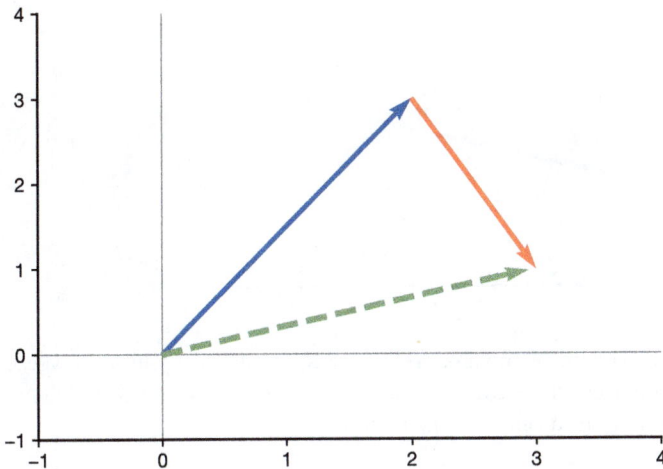

For NumPy vectors, the addition operator + performs vector addition. Let `c` be the vector sum of `a` and `b`. Then:

```
c = a + b
c
```

```
array([3, 1])
```

Since vector addition is just scalar addition for each component, and scalar addition is commutative (is unaffected by order), vector addition is also commutative. We can check this for our example vectors:

```
b + a
```

```
array([3, 1])
```

Compare the dashed line above, representing the result of "chaining" vectors a and b with the plot of a+b shown in the following figure; I set the limits of the axes to be the same as for the chained vectors:

```
import matplotlib.pyplot as plt

plotvecR(a + b)
plt.xlim(-1, 4)
plt.ylim(-1, 4)
```

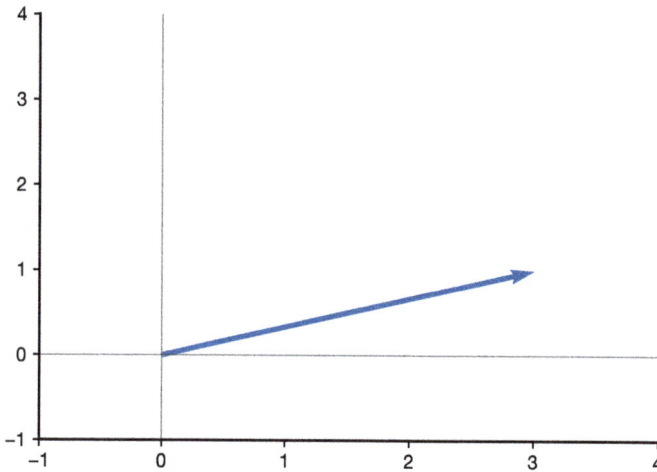

Note that if we change the order of the chained vectors, we still get the same vector from the origin to the head of the second vector. The following figure shows the result when we first draw vector b and then chain a onto the head of b:

```
plotvecR(b, a, newfig=False, chain=True, plotsum=True)
```

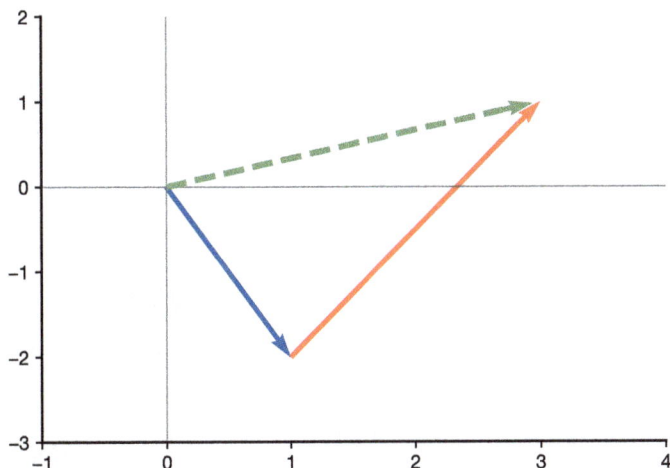

We can sometimes use Python lists in place of NumPy vectors. However, the + operator will not do component-wise addition on Python lists:

```
g = [1, 2]
h = [-3, 4]
print(g + h)
```

```
[1, 2, -3, 4]
```

(The + operator **concatenates** lists.)

2.5.1.1 Properties of Vector Addition

Because vector addition is component-wise scalar addition, it inherits many of its properties from scalar addition:

- *Commutative*: $\mathbf{a} + \mathbf{b} = \mathbf{b} + \mathbf{a}$

- *Associative*: $(\mathbf{a} + \mathbf{b}) + \mathbf{c} = \mathbf{a} + (\mathbf{b} + \mathbf{c})$

- *Identity*: The zero vector is the identity for vector addition: $\mathbf{a} + \mathbf{0} = \mathbf{a}$

2.5.2 Sum of Elements of a Vector

In data science, we often want to sum the data contained in a vector. If $\mathbf{u}$ is a n-vector, then mathematically, the sum of the elements of $\mathbf{u}$ would be written as

$$\sum_{i=0}^{n-1} u_i.$$

(Later in this section, we will introduce another approach for finding the sum using a type of vector-vector multiplication.)

In Python, we can get the sum of the elements in $\mathbf{u}$ using `np.sum()` or the built-in `sum()` method of a NumPy vector:

```
u = np.array( [1, -3, 4, 10] )
np.sum(u)
```

12

```
u.sum()
```

12

2.5.3   Scalar-Vector Multiplication (Scaling)

In scalar-vector multiplication, a vector is multiplied by a scalar (a number). This is achieved by multiplying every component of the vector by the scalar:

DEFINITION

multiplication (scalar-vector)

Given a vector $\mathbf{u}$ and a scalar α, define $\alpha\mathbf{u}$ such that each component i of $\alpha\mathbf{u}$ is given by $(\alpha\mathbf{u})_i = \alpha u_i$. Thus,

$$\alpha\mathbf{u} = \alpha\left[u_0,\ u_1,\ \ldots,\ u_{n-1}\right]^\mathsf{T} = \left[\alpha u_0,\ \alpha u_1,\ \ldots,\ \alpha u_{n-1}\right]^\mathsf{T}.$$

In NumPy, we can multiply a scalar by a vector using the usual $*$ multiplication symbol:

```
a1 = 0.5 * a
print(f'a = {a},    0.5*a = {a1}')
plotvecR(a, a1, labels=['$\mathbf{a}$', '$0.5 \mathbf{a}$'])
```

a = [2 3],   0.5*a = [1. 1.5]

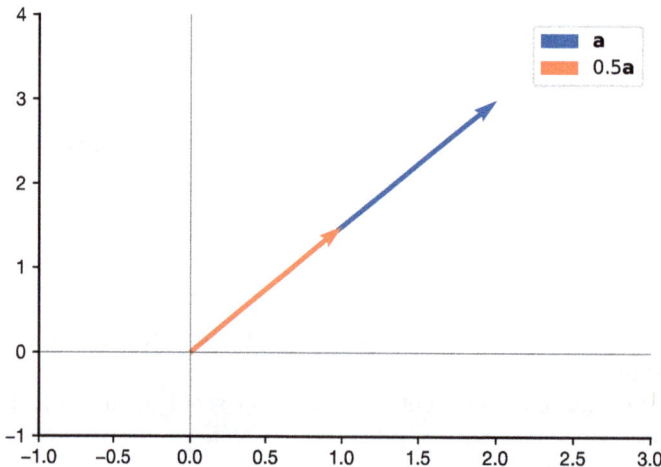

```
b1 = 3 * b
print(f'b = {b},    3*b = {b1}')

# I swapped the order so that the vector b will not
# be hidden by the vector 3b
plotvecR(b1, b, labels=['3$\mathbf{b}$', ' $\mathbf{b}$'])
```

```
b = [ 1 -2],   3*b = [ 3 -6]
```

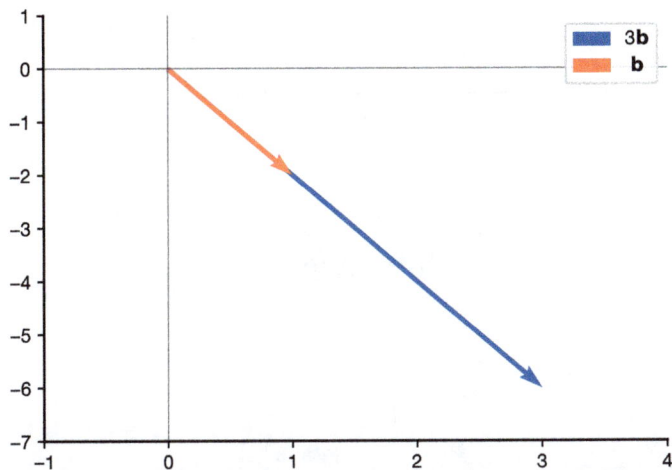

Note that multiplying a vector by a positive scalar yields an output vector that is in the **same direction as the original vector**, but its length has been changed in a way that depends on the value of the scalar. Let's consider the effect of negative scalars:

```
a2 = -0.5 * a
print(f'a = {a},    -0.5*a = {a2}')
plotvecR(a, a2, labels=['$\mathbf{a}$', '$-0.5 \mathbf{a}$'])
```

```
a = [2 3],    -0.5*a = [-1.  -1.5]
```

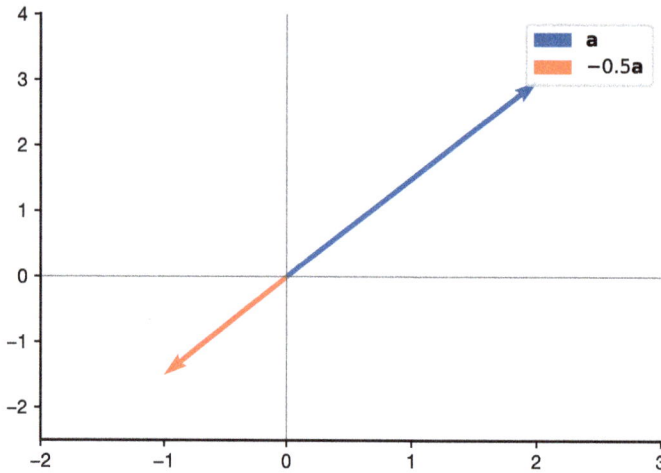

```
b2 = -3 * b
print(f'b = {b},    -3*b = {b2}')
plotvecR(b, b2, labels=['$\mathbf{b}$', '-3$\mathbf{b}$'] )
```

b = [1 -2],   -3*b = [-3 6]

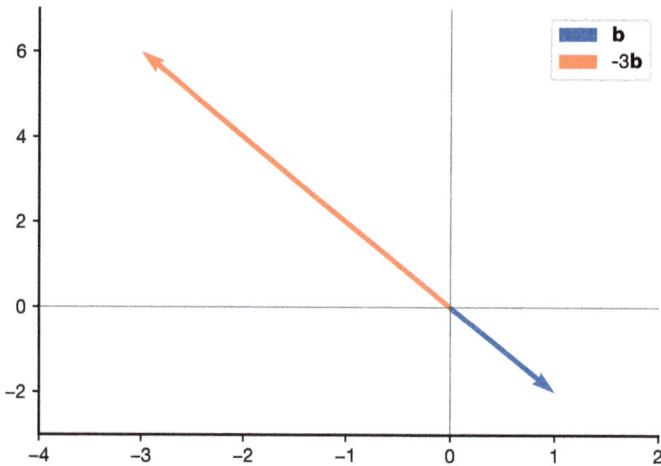

Multiplying by a negative scalar yields a vector that is in the **opposite direction** as the original vector. The length of the new vector is controlled by the scalar's magnitude (i.e., absolute value). For either scalar with magnitude 0.5, the new vector is *shorter* than the original vector. For either scalar with magnitude 3, the new vector is longer than the original vector.

In general, if **u** is a vector and α is a scalar, then:

- if $|\alpha| < 1$, then $\alpha\mathbf{u}$ will be shorter than **u**,

- if $|\alpha| = 1$, then $\alpha\mathbf{u}$ will be the same length as **u**, and

- if $|\alpha| > 1$, then $\alpha\mathbf{u}$ will be longer than **u**.

The details of how to show this have to wait until we define the length of a vector later in this section.

In NumPy, scalar-vector multiplication is considered to be *broadcasting*:

DEFINITION

broadcasting (NumPy)

In NumPy, broadcasting occurs in operations involving two arrays of different shapes. The smaller array is *broadcast* across the larger array by repeating values in the smaller array in such a way that the shapes of the two arrays will match.

In this instance, the scalar is treated as an array with a single element. Under broadcasting, the scalar is repeated to match the size of the vector, and then the two vectors are multiplied component-wise. (This type of multiplication of vectors is called a Hadamard product and is discussed more below.)

Readers interested in learning more about broadcasting can read the NumPy documentation on broadcasting: https://numpy.org/doc/stable/user/basics.broadcasting.html.

Properties of Scalar Multiplication

Since scalar-vector multiplication is component-wise multiplication by a scalar, it inherits the properties below from normal multiplication of real scalars.

If $\mathbf{u}$ and $\mathbf{v}$ are vectors of the same size, and α is a real scalar, then these properties hold:

- *Commutative*: $\alpha\mathbf{u} = \mathbf{u}\alpha$. It does not matter whether the multiplying scalar is on the right or left of the vector.

- *Associative*: If α and β are scalars, then $(\alpha\beta)\mathbf{u} = \alpha(\beta\mathbf{u})$. If multiplying by two scalars, we will get the same result if we do the scalar-scalar multiplication first or the scalar-vector multiplication first.

- *Distributive over scalar addition*: $(\alpha + \beta)\mathbf{u} = \alpha\mathbf{u} + \beta\mathbf{u}$ and $\mathbf{u}(\alpha + \beta) = \mathbf{u}\alpha + \mathbf{u}\beta$

- *Distributive over vector addition*: $\alpha(\mathbf{u} + \mathbf{v}) = \alpha\mathbf{u} + \alpha\mathbf{v}$

2.5.4 Vector Subtraction

We can combine vector addition and scalar-vector multiplication to define vector subtraction. We define $\mathbf{b} - \mathbf{a}$ as $\mathbf{b} + (-1)\mathbf{a}$, which yields

$$\mathbf{b} - \mathbf{a} = \sum_{i=1}^{n-1} (b_i - a_i).$$

If we let $\mathbf{c} = \mathbf{b} - \mathbf{a}$, then we can also write $\mathbf{b} = \mathbf{a} + \mathbf{c}$, so $\mathbf{c}$ is the vector that needs to be added to $\mathbf{a}$ for the result to be $\mathbf{b}$. For example, Fig. 2.2 shows the relations for some example vectors $\mathbf{a}$, $\mathbf{b}$, and $\mathbf{b} - \mathbf{a}$.

The vector $\mathbf{a} - \mathbf{b}$ is the negative of the vector $\mathbf{b} - \mathbf{a}$. Thus both have the same length. The closer that $\mathbf{a}$ and $\mathbf{b}$ are to each other, the smaller the length of the difference will be.

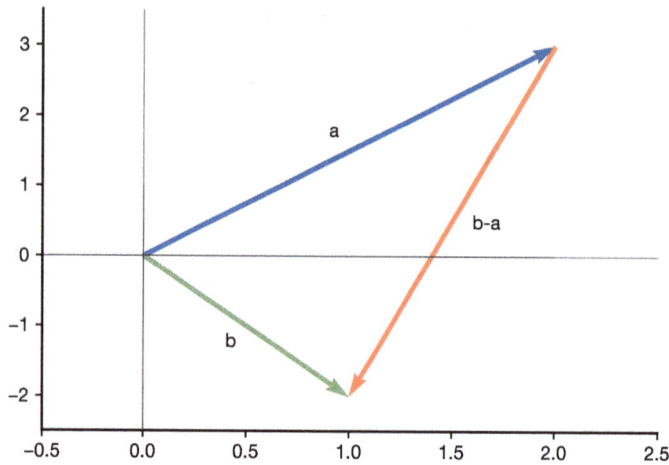

Fig. 2.2: Figure showing the relation among example two-vectors **a** and **b** and the difference vector **b** − **a**.

2.5.5 Component-wise Vector Multiplication: The Hadamard Product

Component-wise multiplication of vectors is also known as the Hadamard or Schur product:

DEFINITION

component-wise multiplication (vectors),
Hadamard product (vectors),
Schur product (vectors)

> Given n-vectors **u** and **v**, the *Hadamard product* or *Schur product* is denoted **u** ⊙ **v** and is the n-vector given by component-wise multiplication of **u** and **v**,
>
> $$\mathbf{u} \odot \mathbf{v} = \begin{bmatrix} u_0 v_0, & u_1 v_1, & \ldots, & u_{n-1} v_{n-1} \end{bmatrix}.$$

Because there are different types of vector multiplication, we need different symbols to distinguish among them. Here we use the symbol ⊙ to indicate component-wise multiplication. (Note that the symbol ∗ that is used for multiplication in many computer languages is not used to indicate multiplication in written mathematics.) When printing from Python, we can use the Unicode character 2299. Here we define a string `odot` containing this Unicode character:

```
odot = '\u2299'
```

Then we can use f-strings to write out a product like **u** ⊙ **y** as

```
print(f'u {odot} v')
```

```
u ⊙ v
```

Component-wise multiplication is relatively uncommon in mathematics; in fact, there is no standard notation for this operation. However, it is a useful building block for other operations and is easy to implement in NumPy. Somewhat confusingly, even though $*$ is not used to indicate multiplication in written mathematics, the standard Python multiplication operator $*$ performs component-wise multiplication:

```
g = np.array( [1, 2] )
h = np.array( [-3, 4] )
print(f'g {odot} h = {g * h}')
```

```
g ⊙ h = [-3  8]
```

Properties of Hadamard Product

Because the Hadamard product is just a collection of pairwise scalar multiplications across all the elements in two vectors, it takes on properties of scalar multiplication, such as being commutative and distributive across addition:

- *Commutative*: $\mathbf{a} \odot \mathbf{b} = \mathbf{b} \odot \mathbf{a}$

- *Associative with scalar multiplication*: $(\gamma \mathbf{a}) \odot \mathbf{b} = \gamma(\mathbf{a} \odot \mathbf{b})$

- *Distributive across vector addition*: $(\mathbf{a} + \mathbf{b}) \odot \mathbf{c} = \mathbf{a} \odot \mathbf{c} + \mathbf{b} \odot \mathbf{c}$

Special cases:

Example 2.14: Hadamard Product with a 0 Vector

The Hadamard product of any vector with the zeros vector is the zeros vector:

```
c= np.array([5,7,9])
z3 = np.zeros(3, dtype=int)

print(f'c {odot} z3   = {c * z3}')
```

```
c ⊙ z3   = [0 0 0]
```

Example 2.15: Hadamard Product with 1s Vector

The Hadamard product of a vector $\mathbf{u}$ with the ones vector returns the vector $\mathbf{u}$:

```
ones3 = np.ones(3, dtype=int)

print(f'c {odot} ones3 = {c * ones3}')
```

```
c ⊙ ones3 = [5 7 9]
```

Example 2.16: Hadamard Product with a Standard Unit Vector

Recall that the standard unit vector e_i is a vector that contains all 0s, except for a single 1 in position i. Thus, the Hadamard product of a vector **u** with e_i consists of all zeros, except it will take the value u_i in position i:

```
e2= np.array([0,1,0])

print(
f'c {odot} e2 = {c * e2}')
```

```
c ⊙ e2 = [0 7 0]
```

Example 2.17: Hadamard Product of a Vector with Itself

If we take the Hadamard product of a vector **u** with itself, element i of the result is simply $u_i \cdot u_i = u_i^2$. Thus, the result is a vector containing the squares of the elements in **u**:

```
print(f'c = {c}')
print(f'c {odot} c = {c * c}')
```

```
c = [5 7 9]
c ⊙ c = [25 49 81]
```

We can also get the squares of the elements by using the ∗∗ operator on a NumPy vector. It will perform element-wise exponentiation:

```
print(f'c ** 2 = {c ** 2}')
```

```
c ** 2 = [25 49 81]
```

2.5.6 Vector-Vector Multiplication: Dot Product

The most common form of multiplication between vectors is called the *inner product* or *dot product*. The input is two vectors of the same length, and the output is a scalar:

DEFINITION

dot product,
inner product (vectors)

> Given n-vectors **u** and **v**, the *dot product* or *inner product* is denoted $\mathbf{u} \cdot \mathbf{v}$ or $\mathbf{u}^\mathsf{T}\mathbf{v}$ and is the **scalar value** given by multiplying corresponding components and summing them up:
>
> $$\mathbf{u} \cdot \mathbf{v} = \sum_{i=0}^{n-1} u_i v_i.$$

Inner product is a concept that can be applied more broadly than to just vectors and can also be denoted using other notation, such as $\langle \mathbf{u}, \mathbf{v} \rangle$.

We use the $\cdot$ symbol (a "dot") for the dot product to distinguish it from component-wise multiplication, which uses $\odot$. To print the "dot" sign in Python, we can use the Unicode character `0xB7`. Let's create a variable called `dot` that contains this unicode value:

```
dot = '\u00B7'
```

The following will print the equivalent of $\mathbf{u} \cdot \mathbf{v}$ in Python:

```
print(f'u{dot}v')
```

```
x·y
```

Example 2.18: Implementing Dot Product in NumPy

The dot product combines two of the operations we previously discussed: component-wise multiplication, followed by summing up the elements. The following code computes the dot product using these two operations:

```
g = np.array( [1, 2] )
h = np.array( [-3, 4] )

gh = g * h

g_dot_h = np.sum(gh)

print(f'g{dot}h  = {g_dot_h}')
```

```
g·h  = 5
```

We can perform the dot product directly using Python's matrix multiply operator, which uses the @ (read "at") symbol.

Example 2.19: Dot Product of NumPy Vectors Using @ Operator

Here is an example of computing the dot product using the @ operator:

```
print(f'g{dot}h  = {g @ h}')
```

```
g·h  = 5
```

2.5.7   Properties of Dot Product

Because dot product is just the Hadamard product followed by a summation operation, it inherits all of the properties of the Hadamard product:

- *Commutative*: $\mathbf{a} \cdot \mathbf{b} = \mathbf{b} \cdot \mathbf{a}$
- *Associative with scalar multiplication*: $(\gamma\mathbf{a}) \cdot \mathbf{b} = \gamma(\mathbf{a} \cdot \mathbf{b})$
- *Distributive across vector addition*: $(\mathbf{a} + \mathbf{b}) \cdot \mathbf{c} = \mathbf{a} \cdot \mathbf{c} + \mathbf{b} \cdot \mathbf{c}$

Special examples:

Example 2.20: Dot Product with 0 Vector

The Hadamard product of any vector with the zeros vector is the zeros vector, so the dot product is the sum over the zeros vector, which is zero:

```
c= np.array([5,7,9])
z3 = np.zeros(3, dtype=int)

print(f'c{dot}z3 = {c @ z3}')
```

```
c·z3 = 0
```

Example 2.21: Inner Product with 1s Vector

The Hadamard product of a vector $\mathbf{u}$ with the ones vector returns $\mathbf{u}$, so the dot product with the ones vector returns the sum of the elements in $\mathbf{u}$:

$$\mathbf{1} \cdot \mathbf{u} = \sum_{i=0}^{n-1} u_i.$$

```
ones3 = np.ones(3, dtype=int)

print(f'c {dot} ones3 = {c @ ones3}')
print(f'sum(c) = {np.sum(c)}')
```

```
c·ones3 = 21
sum(c) = 21
```

Example 2.22: Dot Product with a Standard Unit Vector

The Hadamard product of a vector $\mathbf{u}$ with the standard unit vector $\mathbf{e}_i$ returns a vector of all zeros, except that element i will be u_i. Thus, the dot product is simply u_i:

```
e2 = np.array([0,1,0])

print( f'c{dot}e2 = {c @ e2}')
```

```
c·e2 = 7
```

Example 2.23: Averaging

We can use the dot product to compute the average value of the elements in a n-vector by dotting the vector with a vector whose elements are all $1/n$,

$$\bar{\mathbf{u}} = \left(\frac{1}{n}\right)\mathbf{1} \cdot \mathbf{u} = [1/n, 1/n, \ldots, 1/n] \cdot \mathbf{u}.$$

In the following code, I compute the average using the dot product and compare it with the average computed using the `np.mean()` function:

```
div3 = np.ones(3)/3
print(div3)
```

```
[0.33333333 0.33333333 0.33333333]
```

```
print(f'The dot product of c with a vector of (1/3) values is {div3 @ c:.3f}')
print(f'The average of the values in c using np.mean() is {np.mean(c): .3f}')
```

```
The dot product of c with a vector of (1/3) values is 7.000
The average of the values in c using np.mean() is  7.000
```

Example 2.24: Dot Product of a Vector with Itself: Sum of Squares

Recall that the Hadamard product of a vector $\mathbf{u}$ with itself is a vector of the squares of the elements in $\mathbf{u}$. Then the dot product of a vector with itself is the sum of the squares of the elements in the vector:

$$\mathbf{u} \cdot \mathbf{u} = \sum_{i=0}^{n-1} u_i^2.$$

Let's try this out using our example vector, $\mathbf{c}$:

```
print(f'c = {c}')
print(f'c {odot} c = {c * c}')
print(f'c{dot}c = {c @ c}')
```

```
c = [5 7 9]
c ⊙ c = [25 49 81]
c·c = 155
```

Taking the inner product of a mathematical object with itself is common enough that mathematicians have introduced a special name and notation for it:

DEFINITION

norm squared

For a mathematical object $\mathbf{u}$ with an inner product operator $\langle,\rangle$, the norm squared is denoted by $\|\mathbf{u}\|^2$ and defined as

$$\|\mathbf{u}\|^2 = \langle \mathbf{u}, \mathbf{u} \rangle.$$

For vectors, the inner product operation is the dot product, and the norm squared of a vector $\mathbf{u}$ is $\|\mathbf{u}\|^2 = \mathbf{u} \cdot \mathbf{u}$.

2.5.8 Length or Magnitude of a Vector

Consider again the vector $\mathbf{a} = [2,3]^{\mathsf{T}}$, shown in Fig. 2.3a. Then $\mathbf{a}$ is the hypotenuse of a right triangle with sides 2 and 3, as shown in Fig. 2.3b. Let ℓ_a denote the length of $\mathbf{a}$. By the Pythagorean theorem,

$$\ell_a^2 = 2^2 + 3^2,$$

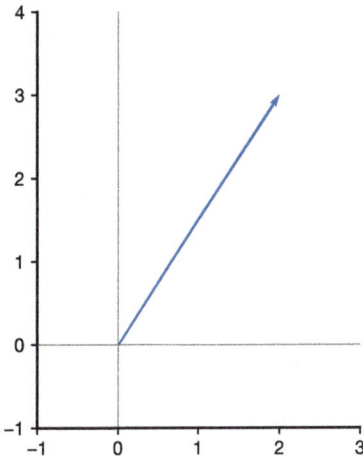

(a) Example vector **a**.

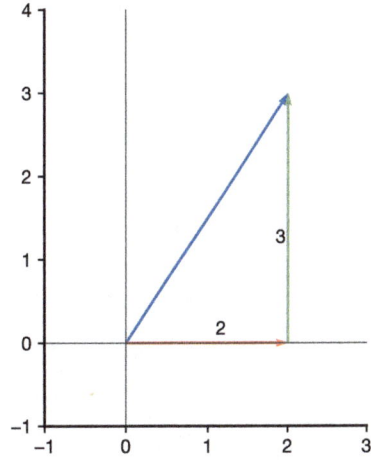

(b) Illustration of vector **a** as hypotenuse of right triangle.

Fig. 2.3: Example vector **a** and its interpretation as the hypotenuse of a right triangle with sides determined by its coordinates.

or

$$\ell_a = \sqrt{2^2 + 3^2}.$$

For any 2-vector $\mathbf{b} = [b_0, b_1]$, the same mathematical approach will give the length ℓ_b as

$$\ell_b = \sqrt{b_0^2 + b_1^2}.$$

The argument inside the square root is simply the norm-squared of **b**, so we can write

$$\ell_b = \sqrt{\|\mathbf{b}\|^2},$$

which we can simplify to

$$\ell_b = \|\mathbf{b}\|.$$

The length of the vector **b** is the norm of **b**. The norm of any mathematical object that has an associated inner product operation is defined below:

DEFINITION

norm

> For a mathematical object **u** with an inner product operator $\langle\,,\,\rangle$, the norm is denoted by $\|\mathbf{u}\|$ and defined as
>
> $$\|\mathbf{u}\| = \sqrt{\langle \mathbf{u}, \mathbf{u} \rangle}.$$

For an n-vector $\mathbf{b}$, the norm is

$$\|\mathbf{b}\| = \sqrt{\mathbf{b} \cdot \mathbf{b}}$$

$$= \sqrt{\sum_{i=0}^{n-1} b_i^2},$$

which is the length of the vector, even if $\mathbf{b}$ has more than two dimensions.

Example 2.25: Norm of Example Vector

Let's start by computing the length of $\mathbf{a}$ by working with the individual elements of $\mathbf{a}$:

```
print(f'|a| = {np.sqrt(a[0]**2 + a[1]**2): .2f}')
```

```
|a| = 3.61
```

Now, let's use the dot product to find the norm of $\mathbf{a}$:

```
print(f'||a|| = {np.sqrt(a @ a): .2f}')
```

```
||a|| = 3.61
```

Finding the norm of a vector is a relatively common operation, so NumPy has a norm operator in the `np.linalg` module:

```
print(f'||a|| = {np.linalg.norm(a): .2f}')
```

```
||a|| = 3.61
```

When using PyTorch tensors to represent vectors, use the `torch.linalg.vector_norm()` function to find the vector norm. However, this function only works on floating point or complex tensors. If you try to use it on a tensor with integer values it will throw an error:

```
import torch
a2 = torch.tensor(a)
torch.linalg.vector_norm(a2)
```

```
---------------------------------------------------------------------
RuntimeError                             Traceback (most recent call last)
Cell In[90], line 3
      1 import torch
      2 a2 = torch.tensor(a)
----> 3 torch.linalg.vector_norm(a2)

RuntimeError: linalg.vector_norm: Expected a floating point or complex tensor
as input. Got Long
```

You can easily convert a PyTorch integer tensor to a float tensor by multiplying it by 1.0 to resolve this issue:

```
torch.linalg.vector_norm(1.0 * a2)
```

```
tensor(3.6056)
```

Example 2.26: Norm of Scaled Vector

Now recall our examples of scaling $\mathbf{a}$ by multiplying it by a constant. Let $\mathbf{w} = \gamma \mathbf{a}$, where γ is some constant. For example, we previously considered $\gamma = 0.5$:

```
a1 = 0.5 * a
plotvec(a, a1, labels=['$\mathbf{a}$', '$0.5 \mathbf{a}$'])
```

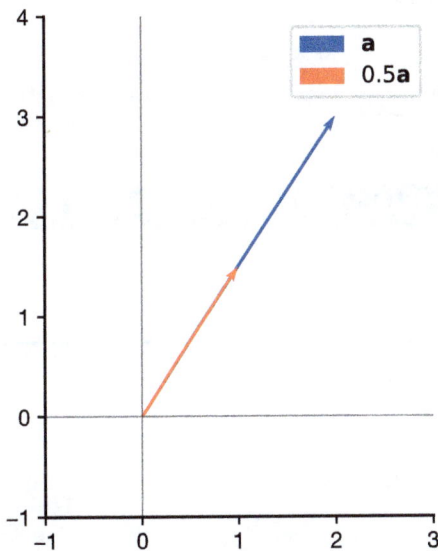

For an arbitrary vector $\mathbf{a}$, we can calculate the length of $\gamma \mathbf{a}$ as

$$\begin{aligned}
\|\gamma\mathbf{a}\| &= \sqrt{\gamma\mathbf{a} \cdot \gamma\mathbf{a}} \\
&= \sqrt{\gamma^2 \mathbf{a} \cdot \mathbf{a}} \\
&= |\gamma| \sqrt{\mathbf{a} \cdot \mathbf{a}} \\
&= |\gamma| \|\mathbf{a}\|.
\end{aligned}$$

For our example, the length of $0.5\mathbf{a}$ is $0.5\|a\|$. Let's check:

```python
w = 0.5*a
print(f'||a|| = {np.linalg.norm(a)}')
print(f'||0.5a|| = {np.linalg.norm(w)}')
```

```
||a|| = 3.605551275463989
||0.5a|| = 1.8027756377319946
```

We can see that the norm of $0.5\mathbf{a}$ is one-half the norm of $\mathbf{a}$.

Normalizing a Vector

Consider what happens if we divide a vector by its norm:

$$\begin{aligned}
\|\tilde{\mathbf{a}}\| &= \left\| \frac{\mathbf{a}}{\|\mathbf{a}\|} \right\| \\
&= \frac{1}{\|\mathbf{a}\|} \|\mathbf{a}\| \\
&= 1,
\end{aligned}$$

where the second line follows from the fact that $1/\|\mathbf{a}\|$ is a constant, so we can factor it out of the norm. We say that $\tilde{\mathbf{a}}$ is a *unit vector*:

> **DEFINITION**
>
> **unit vector**
>
> A vector $\mathbf{v}$ is a *unit vector* if $\|\mathbf{v}\| = 1$.

2.5.9 Distance between Vectors

We define the distance between two n-vectors as follows:

> **DEFINITION**
>
> **distance between vectors**
>
> The *distance* between two n-vectors $\mathbf{a}$ and $\mathbf{b}$ is the norm of the difference between the vectors,
>
> $$d\left(\mathbf{a}, \mathbf{b}\right) = \|\mathbf{a} - \mathbf{b}\| = \|\mathbf{b} - \mathbf{a}\|.$$

Terminology review and self-assessment questions

Interactive flashcards to review the terminology introduced in this section and self-assessment questions are available at la4ds.net/2-5, which can also be accessed using this QR code:

2.6 Vector Correlation and Projection

In this section, I introduce some important math related to representing one vector in terms of another vector. In Chapter 6, we generalize this to the problem of representing a vector given a collection of other vectors.

2.6.1 Vector Correlation

In the field of Statistics, Pearson's correlation is a metric that quantifies the extent that two features can be characterized by a linear relationship. It is a bounded metric that takes a maximum absolute value of 1, when either of the features can be calculated as a linear function of the other. When the sign of Pearson's correlation coefficient is positive, the two features "move together": larger values of one of the features is generally associated with larger values of the other feature. When the sign of Pearson's correlation coefficient is negative, the opposite is true.

We would like to develop a correlation metric for vectors that is similar to Pearson's correlation for data. Ideally, we would like vector correlation to satisfy the following:

- The vector correlation is 1 when the vectors point in the same direction.

- The vector correlation is -1 when the vectors point in the opposite direction.

- The vector correlation is 0 when the vectors are orthogonal in some sense.

To find such a relation, we start with an observation about vectors in $\mathbb{R}^2$. Consider again the vectors $\mathbf{a}$, $\mathbf{b}$, and $\mathbf{b} - \mathbf{a}$ in Fig. 2.2. To simplify the notation, let $\mathbf{c} = \mathbf{b} - \mathbf{a}$. Let θ denote the angle between $\mathbf{a}$ and $\mathbf{b}$. Then the Law of Cosines provides a relation for solving for $\|\mathbf{c}\|^2$:

$$\|\mathbf{c}\|^2 = \|\mathbf{a}\|^2 + \|\mathbf{b}\|^2 - 2\|\mathbf{a}\|\|\mathbf{b}\|\cos\theta. \tag{2.3}$$

We can also use the properties of the norm-squared and dot product to write

$$\begin{aligned}
\|\mathbf{c}\|^2 &= (\mathbf{b} - \mathbf{a}) \cdot (\mathbf{b} - \mathbf{a}) \\
&= \mathbf{b} \cdot \mathbf{b} - \mathbf{b} \cdot \mathbf{a} - \mathbf{a} \cdot \mathbf{b} + \mathbf{a} \cdot \mathbf{a} \\
&= \|\mathbf{a}\|^2 + \|\mathbf{b}\|^2 - 2\mathbf{a} \cdot \mathbf{b}.
\end{aligned} \tag{2.4}$$

Comparing (2.3) and (2.4), we see that

$$\mathbf{a} \cdot \mathbf{b} = \|\mathbf{a}\| \, \|\mathbf{b}\| \cos\theta, \tag{2.5}$$

where θ is the angle between $\mathbf{a}$ and $\mathbf{b}$. (The case where $\theta = 0$ has to be handled separately, but the result is the same.)

Now note the following properties of $\cos\theta$:

- $\cos\theta = 1$ if $\theta = 0$; i.e., if $\mathbf{a}$ and $\mathbf{b}$ point in the exact same direction.

- $\cos\theta = -1$ if $\theta = \pi$; i.e., if $\mathbf{a}$ and $\mathbf{b}$ point in the exact opposite direction.

- $\cos\theta = 0$ if $\theta = \pm\pi/2$; i.e., if $\mathbf{a}$ and $\mathbf{b}$ are orthogonal (perpendicular).

Rewriting (2.5), we have

$$\cos\theta = \frac{\mathbf{a} \cdot \mathbf{b}}{\|\mathbf{a}\| \, \|\mathbf{b}\|}. \tag{2.6}$$

We use this to define a vector correlation metric that holds for any n-vectors:

DEFINITION

correlation (vectors),
cosine similarity

The *correlation* between n-vectors $\mathbf{a}$ and $\mathbf{b}$ is

$$r = \frac{\mathbf{a} \cdot \mathbf{b}}{\|\mathbf{a}\| \, \|\mathbf{b}\|}.$$

It is sometimes called *cosine similarity*.

Using (2.6), we can define the angle between two vectors as shown in the following definition.

DEFINITION

angle between vectors

The angle between n-vectors $\mathbf{a}$ and $\mathbf{b}$ is

$$\theta = \cos^{-1}\left(\frac{\mathbf{a} \cdot \mathbf{b}}{\|\mathbf{a}\| \, \|\mathbf{b}\|}\right).$$

This formula for the angle holds for n-vectors even if $n > 2$.

Note that if the vectors are orthogonal, then $\theta = \pm 90°$. But we can see that $\cos\theta = 0$ occurs if and only if $\mathbf{a} \cdot \mathbf{b} = 0$. Thus, we define *orthogonal vectors* as:

DEFINITION

orthogonal vectors

Vectors $\mathbf{a}$ and $\mathbf{b}$ are *orthogonal* if and only if

$$\mathbf{a} \cdot \mathbf{b} = 0.$$

A special case of orthogonal vectors is if the vectors also have unit norm. Then the vectors are called *orthonormal*:

DEFINITION

orthonormal vectors

Vectors **a** and **b** are *orthonormal* if and only if

$$\mathbf{a} \cdot \mathbf{b} = 0$$

and $\|\mathbf{a}\| = \|\mathbf{b}\| = 1$.

Vector correlation has properties that are similar to those of Pearson's correlation, but the connection is actually much deeper than that. First, let's write vector correlation in terms of the arithmetic operations on the elements of **a** and **b**:

$$r = \frac{\sum_i a_i b_i}{\sqrt{\sum_i a_i^2}\sqrt{\sum_j b_j^2}}.$$

If we treat **a** and **b** as vectors of data features, the Pearson's correlation coefficient is defined as

$$\rho_{ab} = \frac{\text{Cov}(\mathbf{a}, \mathbf{b})}{\sigma_a \sigma_b}$$

$$= \frac{\frac{1}{N-1}\sum_i (a_i - \bar{\mathbf{a}})(b_i - \bar{\mathbf{b}})}{\sqrt{\frac{1}{N-1}\sum_i (a_i - \bar{\mathbf{a}})^2}\sqrt{\frac{1}{N-1}\sum_j (b_j - \bar{\mathbf{b}})^2}}.$$

For the special case $\bar{\mathbf{a}} = \bar{\mathbf{b}} = 0$, this simplifies to

$$\rho_{ab} = \frac{\sum_i a_i b_i}{\sqrt{\sum_i a_i^2}\sqrt{\sum_j b_j^2}}.$$

So, **vector correlation** and **Pearson's correlation** have the same exact form if the elements of each vector average to zero. (This will occur if we subtract off the average of the elements of each vector before calculating the vector correlation.)

2.6.2 Projecting a Vector Onto Another Vector

Consider the problem of taking a vector **b** and writing it as a linear combination of some known vectors $\mathbf{a}_0, \mathbf{a}_1, \mathbf{a}_{m-1}$. Then one of the first goals might be to determine how much of the vector **b** is in the direction of each representation vector $\mathbf{a}_i$.

Consider the example vectors **b** and $\mathbf{a}_i$ shown in Fig. 2.4. If we want to determine how much of **b** is in the direction of $\mathbf{a}_i$, we can draw a vector along $\mathbf{a}_i$ and define the error as the length of the line from the head of that vector to the head of **b**. Three such error lines are shown as the red dotted and solid lines in Fig. 2.5. From inspection, the shortest error line is the one that is orthogonal to $\mathbf{a}_i$, which is shown by the solid line.

Then it is easiest to determine the length of the vector along the direction of $\mathbf{a}_i$ that minimizes the error vector if we rotate the vector $\mathbf{a}_i$ onto the x-axis, as shown in Fig. 2.6.

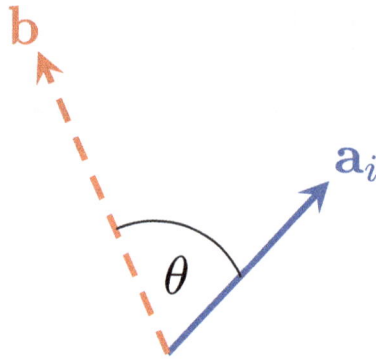

Fig. 2.4: Example vector **b** and a representation vector $\mathbf{a}_i$.

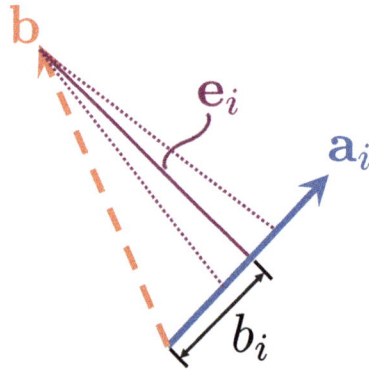

Fig. 2.5: Example vector **b** and a representation vector $\mathbf{a}_i$, along with three error lines.

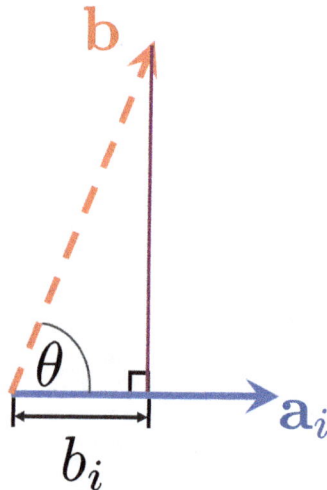

Fig. 2.6: Example vector **b** and a representation vector $\mathbf{a}_i$, rotated to make it easier to see the length of the vector along $\mathbf{a}_i$ that minimizes the error to **b**.

Then from the right triangle, we see $b_i = \|\mathbf{b}\| \cos\theta$. We call this length the *scalar projection* of $\mathbf{b}$ onto $\mathbf{a}_i$. If $|\theta| > 90°$, then the scalar projection will be negative. This occurs if $\mathbf{b}$ needs to be represented in terms of $-\mathbf{a}_i$.

Compare the formula for b_i with $\mathbf{b} \cdot \mathbf{a}_i = \|\mathbf{b}\| \|\mathbf{a}_i\| \cos\theta$. Then we can rewrite the formula for b_i as

$$b_i = \frac{\mathbf{b} \cdot \mathbf{a}_i}{\|\mathbf{a}_i\|}.$$

We use this form in our definition of *scalar projection*:

DEFINITION

scalar projection

Given n-vectors $\mathbf{a}$ and $\mathbf{b}$, the *scalar projection* of $\mathbf{b}$ onto $\mathbf{a}$ has magnitude equal to the length of the vector in the direction of $\mathbf{a}$ that minimizes the error to $\mathbf{b}$. The sign is positive if that component of $\mathbf{b}$ is in the same direction as $\mathbf{a}$ and negative if not. It is given by

$$\frac{\mathbf{b} \cdot \mathbf{a}}{\|\mathbf{a}\|}.$$

We can rewrite the formula for the scalar projection b_i as

$$b_i = \mathbf{b} \cdot \left(\frac{\mathbf{a}_i}{\|\mathbf{a}_i\|} \right).$$

Let $\tilde{\mathbf{a}} = \mathbf{a}/\|\mathbf{a}\|$. Then from Section 2.5.8, $\tilde{\mathbf{a}}$ is a unit vector. Thus, we can also write the scalar projection of $\mathbf{b}$ onto $\mathbf{a}_i$ as $\mathbf{b} \cdot \tilde{\mathbf{a}}_i$.

The *vector projection* is the vector in the direction of $\mathbf{a}_i$ that has length equal to the scalar projection $\mathbf{b} \cdot \tilde{\mathbf{a}}_i$. Since $\tilde{\mathbf{a}}_i$ is a unit vector in the direction of $\mathbf{a}_i$, then the vector projection is simply

$$(\mathbf{b} \cdot \tilde{\mathbf{a}}_i)\,\tilde{\mathbf{a}}_i = \mathbf{b} \cdot \left(\frac{\mathbf{a}_i}{\|\mathbf{a}_i\|} \right) \left(\frac{\mathbf{a}_i}{\|\mathbf{a}_i\|} \right)$$

$$= \frac{\mathbf{b} \cdot \mathbf{a}_i}{\|\mathbf{a}_i\|^2} \mathbf{a}_i.$$

DEFINITION

vector projection

Let $\mathbf{a}$ and $\mathbf{b}$ be n-dimensional vectors. Then the *vector projection* of $\mathbf{b}$ onto $\mathbf{a}$, denoted $\mathrm{proj}_\mathbf{a}\,\mathbf{b}$, is the vector in the direction of $\mathbf{a}$ that minimizes the error to $\mathbf{b}$. It is given by

$$\mathrm{proj}_\mathbf{a}\,\mathbf{b} = \frac{\mathbf{b} \cdot \mathbf{a}}{\|\mathbf{a}\|^2} \mathbf{a}.$$

Example 2.27: Vector Projection in NumPy

Let $\mathbf{a} = [5, 2]^\mathsf{T}$ and $\mathbf{b} = [3, 4]^\mathsf{T}$. Find $\mathrm{proj}_\mathbf{a}\,\mathbf{b}$. Let's start by creating these vectors in NumPy and plotting them using `plotvec()`.

```
import numpy as np
from plotvec import plotvec

a = np.array([5, 2])
b = np.array([3, 4])

plotvec(a, b)
```

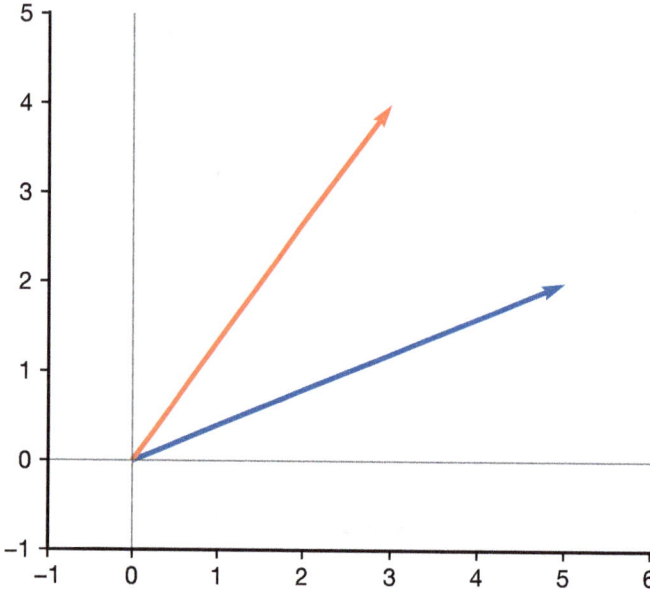

Using (Equation 2.6), we can calculate the angle between these vectors as follows:

```
from numpy.linalg import norm
print(f'{np.rad2deg(np.arccos(a @ b / norm(a) / norm(b))) :.1f} degrees')
```

```
31.3 degrees
```

Let's find and visualize the projection of **b** onto **a**. First, let's find ã; in Python, I will label it `a_t`. Then, let's plot ã on top of the vectors **a** and **b**:

```
import matplotlib.pyplot as plt

a_t = a / norm(a)
plotvec(a, b, alpha=0.7)
plotvec(at, newfig=False, color_offset=2, width=0.015)

plt.annotate('unit $a$', (a_t[0], a_t[1] - 0.25))
```

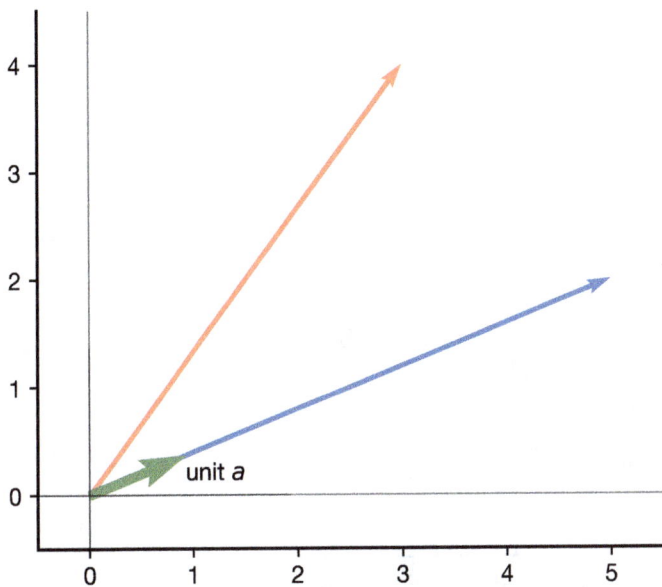

Then the *scalar projection* of **b** onto **ã** is

```
spb = b @ a / norm(a)
spb
```

```
4.270992778072193
```

To get the (vector) projection, $\operatorname{proj}_a \mathbf{b}$, we just need to multiply **ã** by the scalar projection:

```
vpb = spb * a_t
vpb
```

```
array([3.96551724, 1.5862069 ])
```

```
plotvec(a, b, alpha=0.7)
plotvec(vpb, newfig=False, color_offset=2, width=0.02)

plt.annotate('$\operatorname{proj}_\mathbf{a}\mathbf{b}$',
             (vpb[0], vpb[1] - 0.25))
```

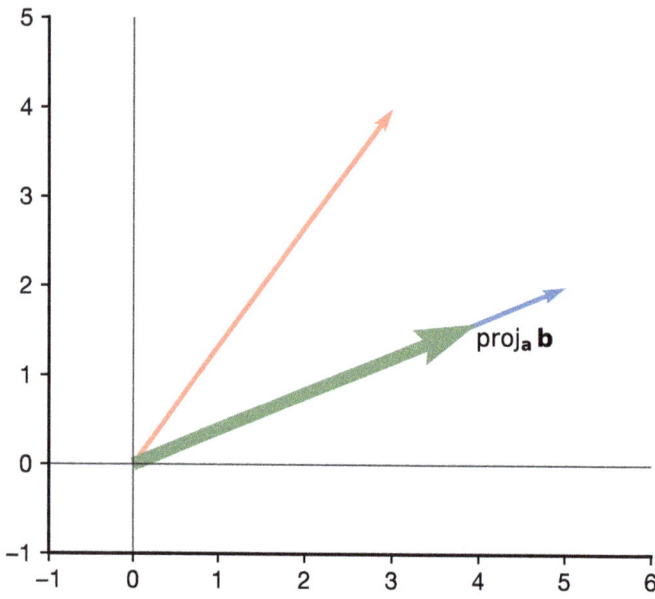

As expected, the error line from the vector projection to **b** is orthogonal to **a**:

```
plotvec(a, b, vpb)
plt.plot([vpb[0], b[0]], [vpb[1], b[1]], color='C4', linestyle='dashed');
```

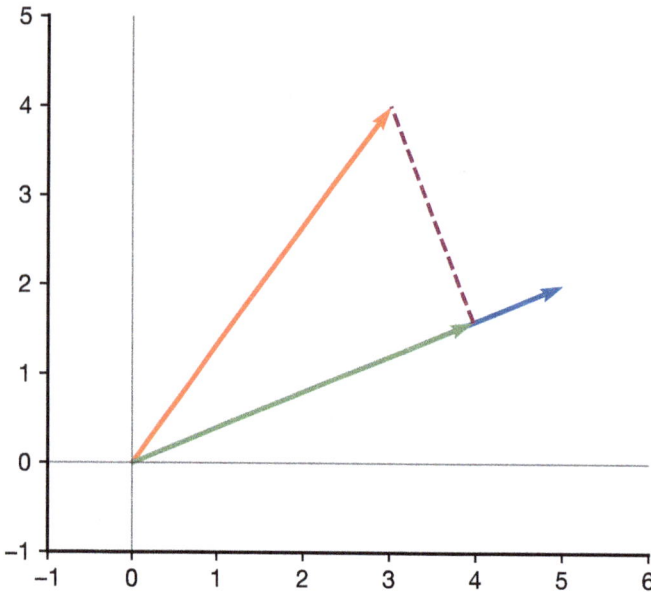

Example 2.28: Vector Projection with Vectors Pointing Away from Each Other

Here is an example vector **g** whose projection onto **a** is in the opposite direction of **a**:

```
a=np.array([5, 2])
g=np.array([-3, -6])

plotvec(a, g)
```

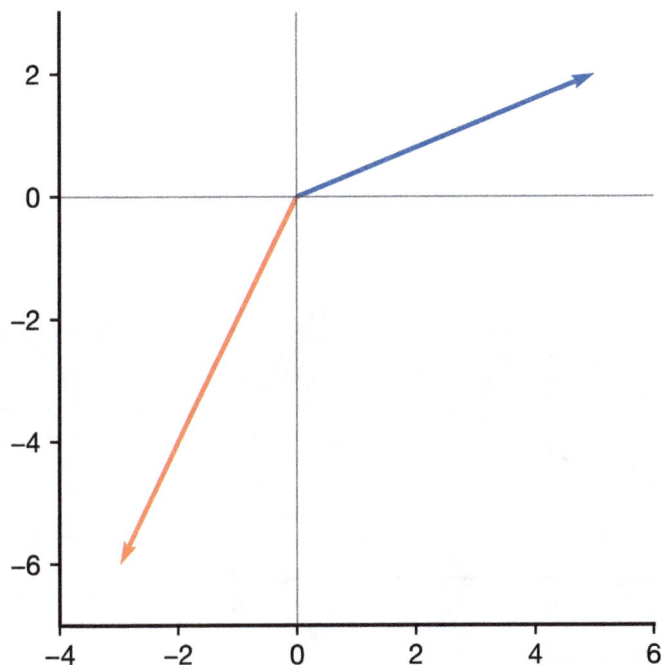

The scalar projection of **g** onto **a** is

```
spg = g @ a / norm(a)
spg
```

```
-5.0137741307804005
```

Since **g** points in the direction of −**a** instead of **a**, the scalar projection is negative. The vector projection $\text{proj}_\mathbf{a}\,\mathbf{g}$ is

```
vpg = spg * a_t
vpg
```

```
array([-4.65517241, -1.86206897])
```

These vectors and the vector projection of **g** onto **a** are shown below. The error line from **g** to the projection of **g** onto **a** is also shown as a dashed line. As expected, the error vector is orthogonal to **a**.

```
plotvec(a, g, vpg)
plt.plot([vpg[0], g[0]], [vpg[1], g[1]], color='C4', linestyle='dashed');
```

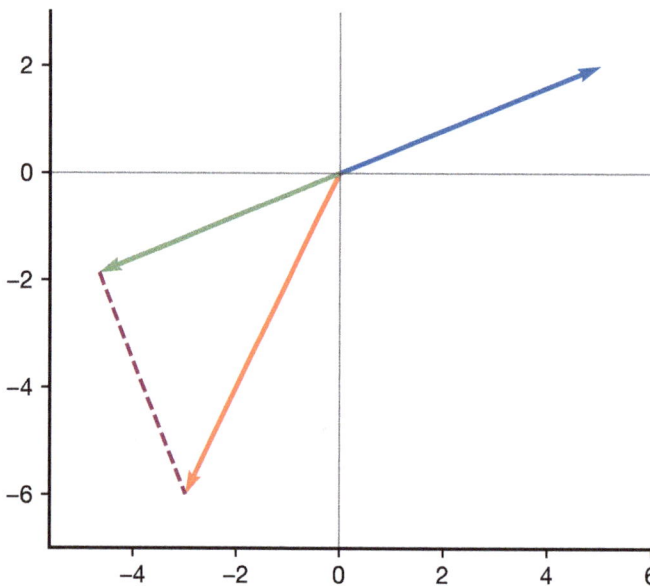

Vector correlation is used to solve the larger problem of representing a vector (or set of vectors) in terms of some other given or derived vectors in Chapter 6.

Terminology review and self-assessment questions

Interactive flashcards to review the terminology introduced in this section and self-assessment questions are available at la4ds.net/2-6, which can also be accessed using this QR code:

2.7 Chapter Summary

In this chapter, I introduced vectors and showed how to visualize them using quiver plots. Several applications for vector data were introduced to show the data science perspective, where vectors are not geometric in nature and instead are used to represent ordered collections of variables or features. In such cases, I showed that vectors may also be illustrated using other types of plots, such as scatter and bar plots. Special vectors were introduced and techniques for creating these vectors in NumPy were given. Then the most common vector operations were introduced and demonstrated using NumPy. Finally, vector correlation, scalar projection, and vector projection were defined. Vector correlation was used to find the representation of a vector in terms of some other given vector, to minimize the norm of the error.

Access a list of key take-aways for this chapter, along with interactive flashcards and quizzes at la4ds.net/2-7, which can also be accessed using this QR code:

3

Matrices and Operations

In Chapter 2, I introduced vectors and their operations. In this chapter, I introduce another important type of mathematical object in linear algebra, *matrices*. If vectors are collections of numbers with a single index (i.e., order 1) that have specific mathematical operations, then matrices are collections of numbers with two indices (order 2) that have their own set of mathematical operations. We will often use matrices to collect multiple vectors and enable simpler and more efficient numerical operations. This chapter focuses on defining matrices, introducing special types of matrices, and defining mathematical operations involving matrices.

3.1 Introduction to Matrices and Tensors

In Chapter 2, we used vectors to store collections of variables or features. However, what if we want to operate on multiple data points or multiple features simultaneously? This will require a mathematical object that can store data in more than one dimension. We will use *tensors* or *matrices* for this purpose:

> **DEFINITION**
>
> **tensor**
>
> A mathematical object that can be represented by an indexed collection of numbers, where the number of indices (i.e., the order) can be greater than 1.

> **DEFINITION**
>
> **matrix**
>
> A tensor of order two; a matrix can be visualized as a two-dimensional array of numbers.

The plural of "matrix" is "matrices". In this book, we will focus our study on matrices. In many data science applications, the numeric data we encounter can be represented as a matrix.

Like a vector, a matrix has *components* or *elements* that can be referenced by their indices. The mathematical notation for a matrix is a two-dimensional table of its elements, enclosed in large square brackets. I will use bold, capital letters for arrays. For example,

$$\mathbf{W} = \begin{bmatrix} 0 & 1 & 2 & 3 \\ 0 & 1 & 4 & 9 \\ 0 & 1 & 8 & 27 \end{bmatrix}.$$

We often describe a matrix in terms of the number of rows and columns it has. A matrix with m rows and n columns is called an $m \times n$ matrix. Throughout this book, in NumPy, and in most mathematical literature, rows are counted or indexed before columns.

Creating Matrices in NumPy

> **Note:**
>
> NumPy has two different data types called `array` (or `ndarray`) and `matrix`, but the `array` type is more general and the one most commonly used in data science and signal processing. For a detailed discussion from the NumPy help pages, see https://numpy.org/devdocs/user/numpy-for-matlab-users.html#array-or-matrix-which-should-i-use.
>
> If you don't want to read that whole section, I will extract a bit of text from the "Short answer" subsection:
>
> "Use arrays."
>
> Thus, in this text, I only use the `array` type, and I will use the terms *array* and *matrix* interchangeably when referring to two-dimensional arrays.

We can represent a matrix using a Numpy `array` by calling `np.array()` with the argument being a list of lists. The outer list contains the rows, and each row is passed as a list containing the components that make up that row. It sounds more confusing than it is in practice!

Example 3.1: Representing a Matrix Using a NumPy Array

Let's make a NumPy version of the matrix above. We will use white space to make the call to `np.array()` more intelligible by putting each row of the matrix on a separate line. Recall that elements in a list must be separated by commas. Then we can create the NumPy array representation of **W** as follows:

```python
import numpy as np

W = np.array([[0, 1, 2, 3],
              [0, 1, 4, 9],
              [0, 1, 8, 27]])
print(W)
```

```
[[ 0  1  2  3]
 [ 0  1  4  9]
 [ 0  1  8 27]]
```

Here, I have put the different rows of the matrix on different rows of the Python code. This is not required but is allowed in Python, and it makes the Python version of the array much easier to interpret. This convention will be used throughout this book.

Matrix and Array Indexing

In mathematical notation, we can refer to an element of a matrix using a lowercase, non-bold form of the array name with a subscript in the form i, j, where i is the row and j is the column. To be consistent with the rest of the text and with NumPy indexing, we use zero-based indexing. For example, $w_{1,2} = 4$. The ith column of the matrix $\mathbf{W}$ is a vector, and we can denote it by $\mathbf{w}_{*i}$. Here, the $*$ symbol is in the row index and can be interpreted as a "wildcard", indicating that we take entries in any row of $\mathbf{W}$ that are in column i. Because matrices are often interpreted as collections of column vectors, the ith column is also denoted $\mathbf{w}_i$, and I use this shorter notation in the rest of the text. The kth row of $\mathbf{W}$ is also a vector, and I denote it by $\mathbf{w}_{i*}$, indicating entries in row i and any column.

Example 3.2: Array Indexing in NumPy

Array indexing in NumPy is performed by putting the indices in square brackets after the variable name for the array. So we can get the element $w_{1,2}$ as

```
print( W[1, 2] )
```

4

If a single index is used in square brackets with an array, it will be interpreted as the **row** index, and all of the columns of that row will be returned as a vector (i.e., the example below returns $\mathbf{w}_{2*}$):

```
print( W[2] )
```

```
[ 0  1  8 27]
```

> **Important!**
>
> Note the inconsistency between mathematical conventions and NumPy/PyTorch conventions:
>
> - Mathematically, for a matrix $\mathbf{W}$, the notation $\mathbf{w}_2$ refers to **column** 2 of $\mathbf{W}$.
>
> - For a NumPy array or PyTorch tensor W, the notation W[2] returns **row** 2 of W.
>
> If you find this confusing, one solution is to always use the other mathematical notation for rows, $\mathbf{w}_{2*}$ and always use the following alternative way of referencing a row in NumPy/PyTorch: W[2,:]. The : entry with no surrounding numbers indicates the entire range of columns.

To access column 3 of W, we can use W[:,3], which indicates to use all rows and column 3. Note the similarity to the alternative mathematical notation for column 3, $\mathbf{w}_{*3}$.

```
print( W[:,3] )
```

```
[ 3  9 27]
```

In general, we can specify a range of rows or columns in the form a:b, where a represents a starting index, and b represents a stopping index (that, as usual, will not be included). In Python, this is called *slicing*. For instance, we can get the components in columns 1 and 2 of row 1 as follows:

```
print( W[1, 1:3] )
```

```
[1 4]
```

If we use a range of the form a:, where the stopping index is omitted, the range goes to the end of that row or column. For example, we can get all the values in column 2 from row 1 to the end as follows:

```
print( W[1:, 2] )
```

```
[4 8]
```

If we use a range of the form `:b`, where the starting index is omitted, then the range starts at the beginning of the row or column. We can retrieve the elements in the first three columns of row 2 as follows:

```
print( W[2, :3] )
```

```
[0 1 8]
```

We can use ranges of rows and columns at the same time to select a subarray. For example,

```
print( W[1:3, 1:3] )
```

```
[[1 4]
 [1 8]]
```

Finally, we can also pass lists of indices to pick out selected elements. For instance, if we want to retrieve the elements at $(0, 3)$ and $(1, 2)$, we could pass two lists of indices: first, the two row indices and, second, the two column indices:

```
print( W[[0,1], [3,2]] )
```

```
[3 4]
```

WARNING

When we create slices of a NumPy array or PyTorch tensor, or when we set a variable equal to an array or tensor, it does **not** create a new array/tensor. It just gives an alternative way to access the original object. This alternative way of accessing the original array/tensor is called a *view*.

Example 3.3: Effects of Changing Elements in a View of an Array

Let's illustrate that slicing a NumPy array creates a view by making a variable V that is the 2×2 submatrix in the upper right-hand corner of W. We can use negative values to index from the end of the rows:

```
V = W[:2, -2:]
print(V)
```

```
[[ 2 -1]
 [-1  9]]
```

Now, let's replace the negative values in V with zeros:

```
V[0,1] = 0
V[1,0] = 0

print(V)
```

```
[[2 0]
 [0 9]]
```

Now let's inspect the values in W:

```
print(W)
```

```
[[ 0  1  2  0]
 [ 0  1  0  9]
 [ 0  1  8 27]]
```

The values in W were updated when we changed the values in V because V was a *view* of W. We will also get a view if we try to create a copy of W by assigning it to a variable Z:

```
Z = W
W[0, 0] = 100
print(Z)
```

```
[[100  1  2   0]
 [  0  1  0   9]
 [  0  1  8  27]]
```

Because W and Z are views into the same NumPy array, changing a value in W also changes that value in Z (and vice versa). To create a variable that is an independent copy of the NumPy array pointed to by W, call the copy() method on the array W when assigning it to a new variable:

```
U = W[:2, :2].copy()
print(U)
```

```
[[100   1]
 [  0   1]]
```

```
U[0,0] = 0
print('U=',U)
print()
print('W=', W)
```

```
U= [[0 1]
 [0 1]]

W= [[100   1   2   0]
 [  0   1   0   9]
 [  0   1   8  27]]
```

For PyTorch tensors, use the `clone()` method to make a copy.

3.1.1   Some Special Types of Matrices

We will often encounter matrices that have the same number of rows and columns; i.e., we have an $m \times m$ matrix. This is called a *square matrix*:

DEFINITION

square matrix

   A matrix for which the number of rows equals the number of columns.

A slice that comes up somewhat frequently is one along the diagonal elements of an $m \times m$ square matrix from element $0,0$ to element $m-1, m-1$. This is called the *main diagonal*:

DEFINITION

main diagonal

   For an $m \times m$ matrix, the *main diagonal* or *principal diagonal* is the vector of m elements at the indices k,k for $k = 0, 1, \ldots, m-1$.

The NumPy function `np.diag()` performs two different operations, depending on the form of its argument:

- When its argument is a two-dimensional array, it returns a vector of the elements on the main diagonal. (More generally, it can be passed another argument to select other diagonals.)

- When its argument is a vector (i.e., a one-dimensional array), it returns a square two-dimensional array that has the elements of the argument along its main diagonal and that has zero for its other elements. The corresponding matrix is called a *diagonal matrix*:

> **DEFINITION**
>
> **diagonal matrix**
>
> A square matrix for which the only nonzero components are along the main diagonal.

Let's illustrate this with some examples. First, I give an example matrix U, and `np.diag(U)` is used to extract the elements on its main diagonal:

```
U = np.array([[1, 2, 3],
              [2, 3, 4],
              [3, 4, 5]])
print(np.diag(U))
```

```
[1 3 5]
```

Now let's create a diagonal matrix:

```
np.diag([7, 9, 11])
```

```
array([[ 7,  0,  0],
       [ 0,  9,  0],
       [ 0,  0, 11]])
```

When using PyTorch to create a diagonal matrix from a vector of its diagonal elements, we can use the `torch.diag()` function, but the input must be a PyTorch tensor:

```
torch.diag( torch.tensor([7, 9, 11]) )
```

```
tensor([[ 7,  0,  0],
        [ 0,  9,  0],
        [ 0,  0, 11]])
```

Some non-diagonal matrices have nonzero components either only in the main diagonal and above or only in the main diagonal and below. These are called *triangular matrices*, and they come in two varieties:

DEFINITION

upper triangular matrix

A square $m \times m$ matrix is an *upper triangular matrix* if its only nonzero elements are in the main diagonal and above; in other words, the only nonzero elements are those at positions k, l that satisfy $l \geq k$.

DEFINITION

lower triangular matrix

A square $m \times m$ matrix is a *lower triangular matrix* if its only nonzero elements are in the main diagonal and below; in other words, the only nonzero elements are those at positions k, l that satisfy $l \leq k$.

DEFINITION

triangular matrix

A square $m \times m$ matrix (two-dimensional array) that is an upper triangular matrix or lower triangular matrix.

To illustrate this, let's start with a non-triangular square NumPy array, V:

```python
V = np.array([[ 1.1, 1.2, 1.3],
              [ 2.1, 2.2, 2.3],
              [ 3.1, 3.2, 3.3]])
print(V)
```

```
[[1.1 1.2 1.3]
 [2.1 2.2 2.3]
 [3.1 3.2 3.3]]
```

We can use the function np.triu() to create an upper triangular matrix with the elements along the main diagonal and above:

```python
print(np.triu(V))
```

```
[[1.1 1.2 1.3]
 [0.  2.2 2.3]
 [0.  0.  3.3]]
```

We can use the function `np.tril()` to create a lower triangular matrix with the elements along the main diagonal and below:

```
print(np.tril(V))
```

```
[[1.1 0.  0. ]
 [2.1 2.2 0. ]
 [3.1 3.2 3.3]]
```

3.1.2 Special Matrices

There are several special matrices that are commonly defined and will simplify our notation later. NumPy offers functions to create versions of these as NumPy arrays. The first of these is the zeros matrix:

DEFINITION

zeros matrix

   A matrix of all zeros.

We will denote the zero matrix of size $m \times n$ by $\mathbf{0}_{m,n}$. For example,

$$\mathbf{0}_{2,5} = \begin{bmatrix} 0 & 0 & 0 & 0 & 0 \\ 0 & 0 & 0 & 0 & 0 \end{bmatrix}.$$

We can create a zeros matrix in NumPy by passing the desired dimensions as a tuple to `np.zeros()`. Thus, we can create a Numpy representation of $\mathbf{0}_{2,5}$ as follows:

```
zeros2_5 = np.zeros( (2,5), dtype=int )
print(zeros2_5)
```

```
[[0 0 0 0 0]
 [0 0 0 0 0]]
```

(Note that in the example above and the following ones, I have set the data type for the matrix elements to `int` so that the output is easier to parse. However, in most cases, it is best to just use the default, which is `float`.)

DEFINITION

ones matrix

   A matrix of all ones.

We will denote the ones matrix of dimension $m \times n$ by $\mathbf{1}_{m,n}$. We can create a ones matrix in NumPy by passing the desired dimensions as a tuple to `np.ones()`. For example, to create a ones matrix with four rows and three columns:

```
np.ones( (4,3), dtype=int )
```

```
array([[1, 1, 1],
       [1, 1, 1],
       [1, 1, 1],
       [1, 1, 1]])
```

Another common matrix that consists only of 0s and 1s is the *identity matrix*:

> **DEFINITION**
>
> **identity matrix**
>
> A diagonal matrix in which the off-diagonal entries are all zero and the di-agonal entries are all equal to 1. The $m \times m$ identity matrix is denoted $\mathbf{I}_m$. If $I_{j,k}$ denotes the entry in row j and column k, then $I_{j,k} = 1$ if $j = k$ and $I_{j,k} = 0$ if $j \neq k$.

We could create an identity matrix in NumPy using `np.ones()` and `np.diag()`. However, the identity matrix is common enough that NumPy provides the command `np.eye()` to create one. When passed a single argument, it will create an identity matrix with the number of rows and columns equal to its argument:

```
I5=np.eye(5, dtype=int)
I5
```

```
array([[1, 0, 0, 0, 0],
       [0, 1, 0, 0, 0],
       [0, 0, 1, 0, 0],
       [0, 0, 0, 1, 0],
       [0, 0, 0, 0, 1]])
```

In the next section, we introduce operations that involve matrices and scalars, vectors, and other matrices.

Terminology review and self-assessment questions

Interactive flashcards to review the terminology introduced in this section and self-assessment questions are available at la4ds.net/3-1, which can also be accessed using this QR code:

3.2   Matrix Operations

There are a wide variety of mathematical operations that are defined between two matrices or between a matrix and either a vector or a scalar. In this section, I introduce many of the most important matrix operations. I leave one of the most important and complicated to its own section: general matrix-matrix multiplication is covered in Section 3.4.

3.2.1 Matrix Addition and Subtraction

One of the simplest matrix operations to define is addition of two matrices of the same dimensions:

DEFINITION

addition (matrices)

For $m \times n$ matrices **A** and **B**, the sum $\mathbf{A} + \mathbf{B}$ is defined as the elementwise sum:

$$\mathbf{A} + \mathbf{B} = \begin{bmatrix} a_{0,0} + b_{0,0} & a_{0,1} + b_{0,1} & \cdots & a_{0,n-1} + b_{0,n-1} \\ a_{1,0} + b_{1,0} & a_{1,1} + b_{1,1} & \cdots & a_{1,n-1} + b_{0,n-1} \\ \vdots & \vdots & \cdots & \vdots \\ a_{m-1,0} + b_{m-1,0} & a_{m-1,1} + b_{m-1,1} & \cdots & a_{m-1,n-1} + b_{m-1,n-1} \end{bmatrix}.$$

Matrix addition in NumPy uses the usual + operator:

```python
import numpy as np

A = np.array([[1, 2],
              [3, 4]])
B = np.array([[1, -1],
              [1, -1]])
print(A + B)
```

```
[[2 1]
 [4 3]]
```

Similarly, matrix subtraction is denoted $\mathbf{A} - \mathbf{B}$ and defined by elementwise subtraction,

$$\mathbf{A} - \mathbf{B} = \begin{bmatrix} a_{0,0} - b_{0,0} & a_{0,1} - b_{0,1} & \cdots & a_{0,n-1} - b_{0,n-1} \\ a_{1,0} - b_{1,0} & a_{1,1} - b_{1,1} & \cdots & a_{1,n-1} - b_{0,n-1} \\ \vdots & \vdots & \cdots & \vdots \\ a_{m-1,0} - b_{m-1,0} & a_{m-1,1} - b_{m-1,1} & \cdots & a_{m-1,n-1} - b_{m-1,n-1} \end{bmatrix}.$$

Matrix subtraction with NumPy arrays uses the usual – operator:

```python
print(A - B)
```

```
[[0 3]
 [2 5]]
```

Matrix addition and subtraction are only formally defined if the dimensions of the two arrays are identical. However, NumPy will allow addition or subtraction between a matrix and an array that matches in one of the corresponding dimensions (i.e., between a $m \times n$ matrix and a $m \times 1$ vector or $1 \times n$ vector) by repeating the vector to match the dimensions of the array via *broadcasting*.

Properties of Matrix Addition

Because matrix addition is component-wise scalar addition, it inherits many of its properties from scalar addition:

- *Commutative* $\mathbf{A} + \mathbf{B} = \mathbf{B} + \mathbf{A}$

- *Associative*: $(\mathbf{A} + \mathbf{B}) + \mathbf{C} = \mathbf{A} + (\mathbf{B} + \mathbf{C})$

- *Identity*: The additive identity is the zero matrix: $\mathbf{A} + \mathbf{0} = \mathbf{0} + \mathbf{A} = \mathbf{A}$.

NumPy/PyTorch Only: Matrix-Scalar Addition and Subtraction

Addition and subtraction between a matrix and a scalar is not usually defined mathematically. However, NumPy will *broadcast* scalar addition or subtraction across all of the elements of a matrix. Thus, if c is a scalar, then $\mathbf{A} + c$ is defined as

$$\mathbf{A} + c = \begin{bmatrix} a_{0,0} + c & a_{0,1} + c & \cdots & a_{0,n-1} + c \\ a_{1,0} + c & a_{1,1} + c & \cdots & a_{1,n-1} + c \\ \vdots & \vdots & \cdots & \vdots \\ a_{m-1,0} + c & a_{m-1,1} + c & \cdots & a_{m-1,n-1} + c \end{bmatrix}.$$

Subtraction is defined similarly. Examples are below:

```
A = np.array([[1, 2],
              [3, 4]])
print(A + 10)
print()
print(A - 1)
```

```
[[11 12]
 [13 14]]

[[0 1]
 [2 3]]
```

3.2.2   Scalar-Matrix Multiplication

A matrix may be left-multiplied or right-multiplied by a scalar with the same result: each element in the matrix is multiplied by the scalar. We can consider this to be broadcasting the scalar multiplication across the elements of the array. Thus,

$$c\mathbf{A} = \mathbf{A}c = \begin{bmatrix} ca_{0,0} & ca_{0,1} & \cdots & ca_{0,n-1} \\ ca_{1,0} & ca_{1,1} & \cdots & ca_{1,n-1} \\ \vdots & \vdots & \cdots & \vdots \\ ca_{m-1,0} & ca_{m-1,1} & \cdots & ca_{m-1,n-1} \end{bmatrix}.$$

Properties of Scalar Multiplication

Since scalar multiplication is component-wise multiplication by a scalar, it inherits the properties below from normal multiplication of real values.

If $\mathbf{A}$ is a vector and c is a real scalar, then these properties hold:

- *Commutative*: $c\mathbf{A} = \mathbf{A}c$. It does not matter whether the multiplying scalar is on the right or left of the vector.

- *Associative*: If b and c are scalars, then $(bc)\mathbf{A} = b(c\mathbf{A})$. If multiplying by two scalars, we will get the same result if we do the scalar-scalar multiplication first or the scalar-matrix multiplication first.

- *Distributive over scalar addition*: $(b + c)\mathbf{A} = b\mathbf{A} + c\mathbf{A}$ and $\mathbf{A}(b + c) = \mathbf{A}b + \mathbf{A}c$.

- *Distributive over matrix addition*: $c(\mathbf{A} + \mathbf{B}) = c\mathbf{A} + c\mathbf{B}$.

3.2.3 Dot Product as Matrix Multiplication

We will build up to general matrix multiplication by starting with the dot product of two n-vectors. Recall from Section 2.5.6 that for n-vectors $\mathbf{u}$ and $\mathbf{v}$,

$$\mathbf{u} \cdot \mathbf{v} = \sum_{i=0}^{n-1} u_i v_i.$$

As mentioned in Section 2.1, vectors are usually written as columns of numbers. When we are working only with vectors and scalars, then we can treat the vectors as one-dimensional objects. However, if we want to perform operations involving matrices and vectors, then we will treat the vectors as occupying one dimension of a two-dimensional matrix. The usual convention, which is followed in this book, is that a vector is a single-column matrix. We call such a vector a *column vector*. For instance, the vector $\mathbf{u}$ can be written as

$$\mathbf{u} = \begin{bmatrix} u_0 \\ u_1 \\ \vdots \\ u_{n-1} \end{bmatrix}.$$

Note also that the dot product of $\mathbf{u}$ and $\mathbf{v}$ can be written in two forms. Until now, we have used the form $\mathbf{u} \cdot \mathbf{v}$, but the dot product can also be written as $\mathbf{u}^\mathsf{T}\mathbf{v}$, which looks like this:

$$\mathbf{u}^\mathsf{T}\mathbf{v} = \begin{bmatrix} u_0 & u_1 & \cdots & u_{n-1} \end{bmatrix} \begin{bmatrix} v_0 \\ v_1 \\ \vdots \\ v_{n-1} \end{bmatrix} \tag{3.1}$$

$$= u_0 v_0 + u_1 v_1 + \ldots + u_{n-1} v_{n-1},$$

where the last line comes from the definition of the dot product.

The vector $\mathbf{u}^\mathsf{T}$ is the transposed version of the column vector $\mathbf{u}$, which results in $\mathbf{u}^\mathsf{T}$ being a *row vector*. We use (3.1) to define multiplication between a row vector and a column vector. It is the same as the dot product: the elements are multiplied component-wise, and the results are summed. In NumPy the @ operator must be used between the two vectors to perform multiplication. For instance, the code below multiplies the row vector $\mathbf{w} = \begin{bmatrix} 1 & 2 & 4 \end{bmatrix}$ by the column vector $\mathbf{z} = \begin{bmatrix} 2 & -2 & 1 \end{bmatrix}^\mathsf{T}$.

```
w = np.array( [[ 1,  2, 4 ]] )
z = np.array( [[ 2, -2, 1 ]] ).T

print(w @ z)
```

[[2]]

3.2.4   Matrix-Vector Multiplication

A matrix can be used to store a collection of n-vectors by storing the vectors as either the rows or columns of the matrix. More generally, given a matrix, we can interpret that matrix as a collection of row or column vectors. In this section, I motivate matrix-vector multiplication by showing how matrix-vector multiplication enables efficient multiplication between a vector and multiple other vectors. Matrix-vector multiplication is usually just indicated by juxtaposing the matrix and vector to be multiplied. For example, in the discussion below, we consider multiplying the matrix $\mathbf{M}$ by the vector $\mathbf{u}$. The product is written $\mathbf{Mu}$.

In all multiplication involving matrices, order is important, and this holds when we have the product of a matrix and a vector. Thus, $\mathbf{Mu} \neq \mathbf{uM}$, and generally one of these products is not even defined. For the product $\mathbf{Mu}$, we say that $\mathbf{u}$ is left-multiplied by $\mathbf{M}$. Matrix-vector multiplication has two different interpretations, each of which will provide useful insights about matrices later. We will refer to these as the *row interpretation* and *column interpretation*.

Row Interpretation of Matrix-Vector Multiplication

If we have a $p \times q$ matrix $\mathbf{M}$ and a $q \times 1$ column vector $\mathbf{u}$, where the kth row of $\mathbf{M}$ is denoted by $\mathbf{m}_{k*}$, then

$$\mathbf{Mu} = \begin{bmatrix} -\!\!-\mathbf{m}_{0*}-\!\!- \\ -\!\!-\mathbf{m}_{1*}-\!\!- \\ \vdots \\ -\!\!-\mathbf{m}_{p-1*}-\!\!- \end{bmatrix} \mathbf{u} = \begin{bmatrix} \mathbf{m}_{0*} \cdot \mathbf{u} \\ \mathbf{m}_{1*} \cdot \mathbf{u} \\ \vdots \\ \mathbf{m}_{p-1*} \cdot \mathbf{u} \end{bmatrix}.$$

Here, the horizontal lines are shown to help indicate that $\mathbf{m}_{k*}$ is a row vector.

The ith component of the result vector is the dot product of the ith row of $\mathbf{M}$ with the vector $\mathbf{u}$. Each element of the output vector is a linear combination of the elements in $\mathbf{u}$, and we say that left-multiplication by $\mathbf{M}$ is a *linear transformation* applied to $\mathbf{u}$.

Computation of Matrix-Vector Product by Hand

To compute a matrix-vector product by hand, the usual approach is to start at the top of the matrix and iterate down the rows. For each row, we compute the dot product with $\mathbf{u}$, which means that we just multiply the elements in the row by the elements in $\mathbf{u}$ and sum.

Let's illustrate this using a concrete example. Consider the product

$$\begin{bmatrix} 3 & 4 \\ -1 & 2 \\ 2 & 3 \end{bmatrix} \begin{bmatrix} 2 \\ -1 \end{bmatrix}.$$

Let the result of product be denoted by **z**. From our previous work, we know the product of a 3×2 matrix and 2×1 is a 3×1 vector. Each row of the input matrix results in one entry in the output vector, and each entry in the output vector depends on only one row in the input matrix.

Starting with the first row of the matrix, we compute the dot product of the row with the column vector as follows: work simultaneously across the row of the matrix and down the column of the vector, computing the products of the corresponding elements, and then sum all those products. The first two elements to be multiplied are shown below:

$$\begin{bmatrix} 3 & 4 \\ -1 & 2 \\ 2 & 3 \end{bmatrix} \begin{bmatrix} 2 \\ -1 \end{bmatrix} = \begin{bmatrix} z_0 \\ z_1 \\ z_3 \end{bmatrix}$$

$$z_0 = 3(2) + \ldots$$

Then the next two elements are multiplied and added to the first element to get the first element of the output vector:

$$\begin{bmatrix} 3 & 4 \\ -1 & 2 \\ 2 & 3 \end{bmatrix} \begin{bmatrix} 2 \\ -1 \end{bmatrix} = \begin{bmatrix} z_0 \\ z_1 \\ z_3 \end{bmatrix}$$

$$z_0 = 3(2) + 4(-1) = 2$$

Proceeding to the second row, we simultaneously proceed across the elements in the row and down the elements in the vector, compute the products, and then sum them:

$$\begin{bmatrix} 3 & 4 \\ -1 & 2 \\ 2 & 3 \end{bmatrix} \begin{bmatrix} 2 \\ -1 \end{bmatrix} = \begin{bmatrix} 2 \\ z_1 \\ z_3 \end{bmatrix}$$

$$z_1 = -1(2) + 2(-1) = -4$$

Finally, we conduct the same procedure using the last row of the matrix:

$$\begin{bmatrix} 3 & 4 \\ -1 & 2 \\ 2 & 3 \end{bmatrix} \begin{bmatrix} 2 \\ -1 \end{bmatrix} = \begin{bmatrix} 2 \\ -4 \\ z_3 \end{bmatrix}$$

$$z_2 = 2(2) + 3(-1) = 1$$

The final result follows:

$$\begin{bmatrix} 3 & 4 \\ -1 & 2 \\ 2 & 3 \end{bmatrix} \begin{bmatrix} 2 \\ -1 \end{bmatrix} = \begin{bmatrix} 2 \\ -4 \\ 1 \end{bmatrix}$$

Let's check our work using NumPy. As with vector-vector multiplication, we use the @ sign for matrix-vector multiplication:

```
M = np.array([[3, 4],
              [-1, 2],
              [2, 3]])
u = np.array([[2],
              [-1]])
print(M @ u)
```

```
[[ 2]
 [-4]
 [ 1]]
```

Note that I explicitly made the vector u to be a column vector, and the result is also a column vector. If you instead use a one-dimensional form for the vector, NumPy will compute the product in the same way, but the result will be returned as a one-dimensional vector:

```
u2 = np.array([2, -1])
print(M @ u2)
```

```
[ 2 -4  1]
```

The procedure described above for calculating the matrix-vector product generalizes to any size matrix and any size vector, provided that the number of columns of the matrix is equal to the number of rows (i.e., the size) of the column vector.

Column Interpretation of Matrix-Vector Multiplication

There is another interpretation of matrix-vector multiplication that is useful. Consider the form of the output vector:

$$\begin{bmatrix} m_{0,0}u_0 + m_{0,1}u_1 + \ldots + m_{0,q-1}u_{q-1} \\ m_{1,0}u_0 + m_{1,1}u_1 + \ldots + m_{1,q-1}u_{q-1} \\ \vdots \\ m_{p-1,0}u_0 + m_{p-1,1}u_1 + \ldots + m_{p-1,q-1}u_{q-1} \end{bmatrix}$$

Look at the first term in each of the summations. If we collect all those terms into a vector, we have the column vector $u_0\mathbf{m}_0$. If we do that for each of the terms in the summation, we can rewrite the product as

$$\begin{bmatrix} u_0\mathbf{m}_0 + u_1\mathbf{m}_1 + \ldots + u_{q-1}\mathbf{m}_{q-1} \end{bmatrix}.$$

Thus, the result vector is a **linear combination of the columns** of **M**, where the coefficients are the values in the corresponding positions in the vector **u**.

Let's confirm that we get the same answer for the matrix-vector product **Mu** by implementing this approach in Python. Recall that the ith column of the NumPy array M is M{[:,i]}. Then using the column interpretation of matrix-vector multiplication, the product **Mu** is

```
u[0]*M[:,0] + u[1]*M[:,1]
```

```
array([ 2, -4,  1])
```

To find the matrix-vector product **Mu** using the column interpretation by hand, first write the product as a linear combination of the columns of **M**, where the coefficients are the corresponding entries in **u**. For our example, this is the result:

$$\begin{bmatrix} 3 & 4 \\ -1 & 2 \\ 2 & 3 \end{bmatrix} \begin{bmatrix} 2 \\ -1 \end{bmatrix} = 2 \begin{bmatrix} 3 \\ -1 \\ 2 \end{bmatrix} + (-1) \begin{bmatrix} 4 \\ 2 \\ 3 \end{bmatrix}.$$

Next, carry out all the scalar-vector multiplications, and finally add the scaled vectors:

$$= \begin{bmatrix} 6 \\ -2 \\ 4 \end{bmatrix} + \begin{bmatrix} -4 \\ -2 \\ -3 \end{bmatrix}$$

$$= \begin{bmatrix} 2 \\ -4 \\ 1 \end{bmatrix}.$$

3.2.5 Application of Matrix-Vector Multiplication: Feature Extraction

Let's show how matrix multiplication can be used for *feature extraction*:

> **DEFINITION**
>
> **feature extraction**
>
> The process of creating new features from raw data, often with the intent of reducing the number of features.

We will show how to perform feature extraction using the Iris data set, which contains data about flowers from three different Iris species. This is one of the oldest and most famous data sets for classification problems (where the goal is to infer the correct class from a data point's features). This is a relatively simple data set, and we are going to simplify it more for the purposes of this section. The data set is from Robert Fisher's paper "The use of multiple measurements in taxonomic problems", *Annual Eugenics*, 7, Part II, 179–188 (1936).

The Iris data set can be loaded from a Python library called `scikit-learn`, which contains many data sets and tools for machine learning. We can load the data set as follows:

```python
from sklearn import datasets

iris = datasets.load_iris()
```

According to the `DESCR` property of the `scikit-learn` iris data set object:

> This is perhaps the best known database [sic] to be found in the
> pattern recognition literature. Fisher's paper is a classic
> in the field and is referenced frequently to this day.... The
> data set contains 3 classes of 50 instances each, where each
> class refers to a type of iris plant.

(To see the full description, run `print(iris['DESCR'])` after loading the Iris data set.) The
`DESCR` property also explains the features present in the data set:

```
print(iris['DESCR'][:500])
```

```
.. _iris_dataset:

Iris plants dataset
--------------------

**Data Set Characteristics:**

    :Number of Instances: 150 (50 in each of three classes)
    :Number of Attributes: 4 numeric, predictive attributes and the class
    :Attribute Information:
        - sepal length in cm
        - sepal width in cm
        - petal length in cm
        - petal width in cm
        - class:
                - Iris-Setosa
                - Iris-Versicolour
                - Iris-Virginica
```

As the description indicates, each data point contains four features, which are labeled
in `iris['feature_names']`. The data itself is contained in `iris['data']`:

```
print(iris['feature_names'])
print(iris['data'][:5])
```

```
['sepal length (cm)', 'sepal width (cm)',
 'petal length (cm)', 'petal width (cm)']
[[5.1 3.5 1.4 0.2]
 [4.9 3.  1.4 0.2]
 [4.7 3.2 1.3 0.2]
 [4.6 3.1 1.5 0.2]
 [5.  3.6 1.4 0.2]]
```

(An Iris flower consists of similarly colored sepals and petals, but the sepals are longer and have a bulb shape that is wider than the petals, as is indicated by the data.)

Each data point is also associated with its correct classification or classification *target*. The `iris['target']` member contains the numerical classification target, and `iris['target_names']` contains the description of each class, which in this case are three different types of Irises:

```
iris['target'], iris['target_names']
```

```
(array([0, 0, 0, 0, 0, 0, 0, 0, 0, 0, 0, 0, 0, 0, 0, 0, 0, 0, 0, 0, 0,
        0, 0, 0, 0, 0, 0, 0, 0, 0, 0, 0, 0, 0, 0, 0, 0, 0, 0, 0, 0, 0,
        0, 0, 0, 0, 0, 0, 1, 1, 1, 1, 1, 1, 1, 1, 1, 1, 1, 1, 1, 1, 1,
        1, 1, 1, 1, 1, 1, 1, 1, 1, 1, 1, 1, 1, 1, 1, 1, 1, 1, 1, 1, 1,
        1, 1, 1, 1, 1, 1, 1, 1, 1, 1, 1, 1, 2, 2, 2, 2, 2, 2, 2, 2, 2,
        2, 2, 2, 2, 2, 2, 2, 2, 2, 2, 2, 2, 2, 2, 2, 2, 2, 2, 2, 2, 2,
        2, 2, 2, 2, 2, 2, 2, 2, 2, 2, 2, 2, 2, 2, 2, 2, 2, 2, 2, 2]),
 array(['setosa', 'versicolor', 'virginica'], dtype='<U10'))
```

The first 50 elements are of class 0, the next 50 are of class 1, and the final 50 are of class 2. We will use this fact when plotting the data.

The usual goal when working with this data set is to determine a classification function that maps from the four-dimensional data to the three classes. Here, we simplify the problem to just the first two classes and the first two features:

```
class01 = np.where(iris['target']<2)[0]

R = iris.data[class01][:,:2]
target2 = iris.target[class01]
```

The reduction to two features (sepal length and sepal width) allows us to plot the data as points using a scatter plot, as shown in Fig. 3.1. The code to generate this figure is online at la4ds.net/3-2.

The plot of the data in Fig. 3.1 shows that these two features are almost sufficient to distinguish between these two classes: data in the lower-right of the plot correspond to Versicolor, whereas data in the upper left correspond to Setosa. In fact, if we rotated the data by 40° counter-clockwise, we could distinguish between the classes using only the first feature. This feature can be created using by projecting each data point onto the vector $\tilde{\mathbf{x}} = [\cos 40°, -\sin 40°]^{\mathsf{T}}$. I am going to create that as a 2×1 column vector in NumPy:

```
x40cw = np.array([[ 0.76604444],
                  [-0.64278761]])
```

If we let the data point i be the **column** vector $\mathbf{d}_i$, then the $\mathbf{R}$ matrix has rows that are the *transpose* of these data vectors:

$$\mathbf{R} = \begin{bmatrix} \mathbf{d}_0^{\mathsf{T}} \\ \mathbf{d}_1^{\mathsf{T}} \\ \vdots \\ \mathbf{d}_{99}^{\mathsf{T}} \end{bmatrix}.$$

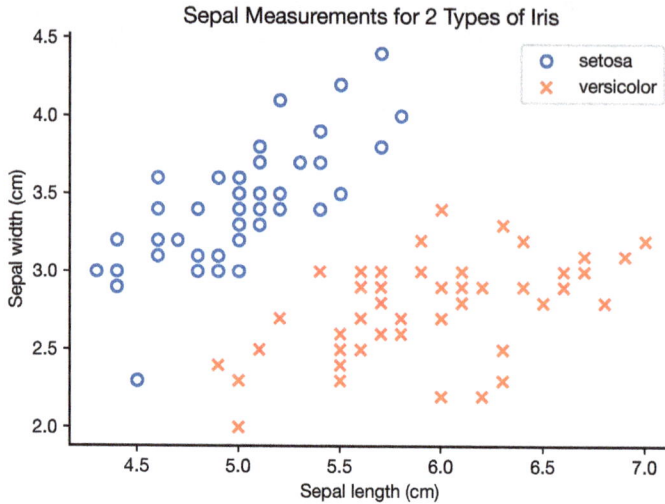

Fig. 3.1: Plot of first two classes and first two features of Iris data set.

We can choose any row of R with indexing and perform feature extraction by carrying out matrix multiplication between that 1×2 row $\mathbf{d}_i^\mathsf{T}$ and the 2×1 column vector $\mathbf{x}$. For instance, here is row 2 times the feature extraction vector:

```
R[2] @ x40cw
```

```
array([1.54348852])
```

The power of matrix multiplication is that it can do all of these row-times-vector multiplications with a single operation. If we multiply the 100×2 matrix $\mathbf{R}$ by the 2×1 feature-extraction vector $\tilde{\mathbf{x}}$, then we get a 100×1 column vector:

$$\mathbf{R}\,\tilde{\mathbf{x}} = \begin{bmatrix} -\mathbf{d}_0^\mathsf{T}- \\ -\mathbf{d}_1^\mathsf{T}- \\ \vdots \\ -\mathbf{d}_{99}^\mathsf{T}- \end{bmatrix} \tilde{\mathbf{x}} = \begin{bmatrix} \mathbf{d}_0^\mathsf{T}\tilde{\mathbf{x}} \\ \mathbf{d}_1^\mathsf{T}\tilde{\mathbf{x}} \\ \vdots \\ \mathbf{d}_{99}^\mathsf{T}\tilde{\mathbf{x}} \end{bmatrix}.$$

Here, the horizontal lines are again used to indicate that the $\mathbf{d}_i^\mathsf{T}$ are row vectors. Thus, we can perform the feature extraction with just the single command:

```
new_feature = R @ x40cw
```

For reference, let's print out the shapes of these three vectors. Here I am showing them in the form that allows matching up with the equation

$$\mathbf{R}\,\tilde{\mathbf{x}} = \mathbf{f},$$

where $\mathbf{f}$ is the vector containing the extracted features:

```
print( f'{R.shape} x {x40cw.shape} = {new_feature.shape}')
```

```
(100, 2) x (2, 1) = (100, 1)
```

The inner two dimensions have to agree for the dot product to work, and these two dimensions are reduced to a single value for each of the other dimensions. Thus, the dimension of the result is determined by the outer two dimensions, which, in this case, is $(100, 1)$. In fact, this result holds for all multiplication involving matrices. If $\mathbf{A}$ has dimensions $k \times \ell$ and $\mathbf{B}$ has dimensions $\ell \times m$, then $\mathbf{AB}$ is a matrix with dimensions $k \times m$.

Now let's visualize the data to make sure that we achieved our goal of feature extraction. The following code plots the values of the new feature and shows each class. I have used a random model to choose the y-values in the graph because otherwise too many of the data points are overlapping and hard to see.

```
import scipy.stats as stats
import matplotlib.pyplot as plt

# use random values for the y data so the
# individual points are easier to see
N = stats.norm(0.6, 0.05)
ypos = N.rvs(100)


plt.figure(figsize=(6,3))

# Plot the remaining points
plt.scatter(new_feature[:50], ypos[:50],
            color='C0', alpha=0.6,
            marker='o', facecolor='none')
plt.scatter(new_feature[50:-1], ypos[50:-1],
            color='C1', alpha=0.8,
            marker='x')

plt.xlabel('Feature');
plt.ylim(0.2,1.2);
plt.yticks([]);
plt.gca().spines['left'].set_visible(False)
```

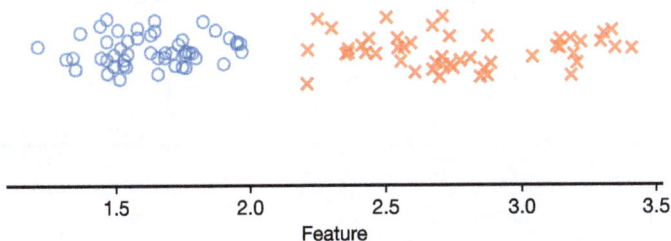

Let's study this type of matrix-vector multiplication a bit more to build up some additional knowledge about it. Suppose we let $\mathbf{D}$ be the matrix whose *columns* are the data vectors,

$$\mathbf{D} = \begin{bmatrix} | & | & & | \\ \mathbf{d}_0 & \mathbf{d}_1 & \cdots & \mathbf{d}_{99} \\ | & | & & | \end{bmatrix},$$

where I have added the vertical bars to help convey the sense that each of the $\mathbf{d}_i$ is a column vector. Then the matrix-vector product $\mathbf{D}^{\mathsf{T}}\tilde{\mathbf{x}}$ looks very similar to the dot product $\mathbf{d}_i^{\mathsf{T}}\tilde{\mathbf{x}}$ and is equal to the vector of dot products of the columns of $\mathbf{D}$ with the vector $\tilde{\mathbf{x}}$:

$$\mathbf{D}^{\mathsf{T}}\tilde{\mathbf{x}} = \begin{bmatrix} \mathbf{d}_0^{\mathsf{T}}\tilde{\mathbf{x}} \\ \mathbf{d}_1^{\mathsf{T}}\tilde{\mathbf{x}} \\ \vdots \\ \mathbf{d}_{99}^{\mathsf{T}}\tilde{\mathbf{x}} \end{bmatrix}.$$

Terminology review and self-assessment questions

Interactive flashcards to review the terminology introduced in this section and self-assessment questions are available at la4ds.net/3-2, which can also be accessed using this QR code:

3.3 Matrix-Vector Multiplication as a Linear Transformation

We start by defining a *vector space*:

DEFINITION

vector space

> A *vector space* consists of a set of vectors and scalars, along with addition and multiplication operators, such that the result of any scalar-vector multiplication and any addition of vectors in the vector space results in a vector contained in that vector space.

We say that a vector space is *closed* under scalar multiplication and vector addition. In this book, the multiplication and addition operators will always be the standard operators for real vectors and scalars and will not be explicitly listed when discussing vector spaces. Until Chapter 6, we only consider Euclidean vector spaces:

DEFINITION

Euclidean vector space

> A *Euclidean vector space* of dimension n is denoted $\mathbb{R}^n$ and contains all real n-vectors.

Consider an n-vector $\mathbf{u}$ and an $m \times n$ matrix $\mathbf{M}$. Then $\mathbf{v} = \mathbf{Mu}$ is an m-vector, where each component of $\mathbf{v}$ is a linear combination of the components of $\mathbf{u}$. Since we can do this for any $\mathbf{u} \in \mathbb{R}^n$ and for each $\mathbf{u}$, $\mathbf{Mu} \in \mathbb{R}^m$, we call this a *linear transformation*:

> **DEFINITION**
>
> ### linear transformation (from $\mathbf{R}^n$ to $\mathbf{R}^m$)
>
> Let $\mathbf{M}$ be an $m \times n$ real matrix. For any vector $\mathbf{u} \in \mathbb{R}^n$, $\mathbf{v} = \mathbf{M}\mathbf{u}$ has components that are a linear combination of the components of $\mathbf{u}$, and $\mathbf{v} \in \mathbb{R}^m$. We say that $\mathbf{M}$ is a *linear transformation* from $\mathbb{R}^n$ to $\mathbb{R}^m$.

One of the most common linear transformations is from a vector space to the same vector space; i.e., if M is a $n \times n$ square matrix, then both the input and output vectors belong to $\mathbb{R}^n$. However, the output vector can have a completely different length and point in a completely different direction than the input vector. Moreover, we will see that the effect of multiplying by $\mathbf{M}$ is different depending on the direction in which the vector $\mathbf{u}$ is pointing. On the other hand, it doesn't really depend on the length $\|\mathbf{u}\|$ because we can write the vector $\mathbf{u}$ as $\mathbf{u} = \|\mathbf{u}\|\tilde{\mathbf{u}}$, where $\tilde{\mathbf{u}}$ is a unit vector. Then

$$\mathbf{M}\mathbf{u} = \mathbf{M}\,\|\mathbf{u}\|\,\tilde{\mathbf{u}}$$
$$= \|\mathbf{u}\|\,(\mathbf{M}\,\tilde{\mathbf{u}}),$$

so different lengths $\|\mathbf{u}\|$ just change the result proportionately. Thus, it is sufficient to understand how $\mathbf{M}$ affects different vectors by studying its effect on the unit vectors.

Let's visualize this effect in 2-D space for the following matrix:

```
M = np.array([[0.5, -4],
              [-2,   3]])
```

We will use the function `transform_unit_vecs()` from the PlotVec library to visualize the effect of this matrix on unit vectors. By default, this function creates 16 unit vectors, evenly spaced around the unit circle, as input vectors. It then left-multiplies each of these vectors by the specified matrix and calculates the 16 output vectors. The input vectors are plotted on the left, and the output vectors are plotted on the right:

```
from plotvec import transform_unit_vecs
transform_unit_vecs(M)
```

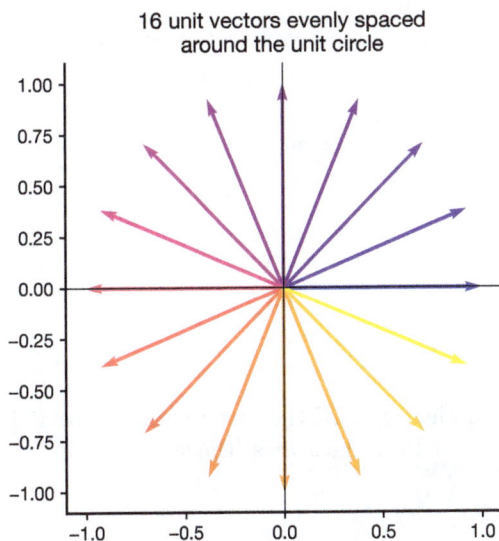

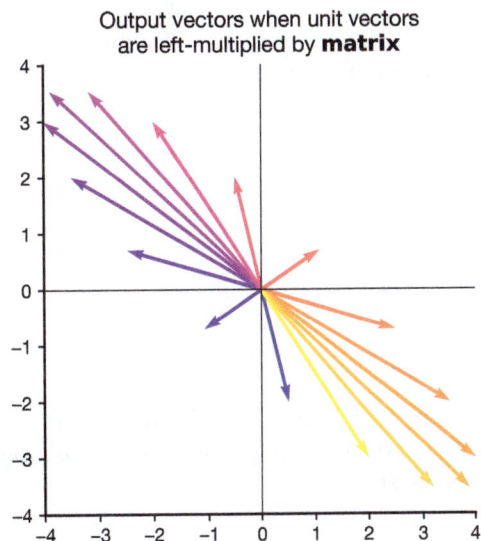

The result is that the original unit vectors experience a combination of rotation, stretching, and flipping (since the order of the colors around the circle is reversed, and that cannot be achieved with just a rotation). Let's investigate this in more detail by considering a couple of examples.

Consider first the unit vector $\mathbf{u}_0 = [1, 0]^\mathsf{T}$, which lies on the x-axis. The corresponding output vector is

```
u0= np.array([1,0])

v0 = M @ u0
v0
```

```
array([ 0.5, -2. ])
```

Since the input vector lies on the positive x-axis, the angle of rotation is equal to the angle of the output vector from the x-axis. This angle can easily be calculated (in degrees) using trigonometry as

```
np.rad2deg(np.arctan2(v0[1], v0[0]))
```

```
-75.96375653207353
```

The input and output vectors are shown in Fig. 3.2. Although the input vector's length is 1, the output vector's length is

```
np.linalg.norm(v0)
```

```
2.0615528128088303
```

If we instead consider the input vector $\mathbf{u}_1 = [0, 1]^\mathsf{T}$, which corresponds to a unit vector on the positive y-axis, the output is

```
u1= np.array([0,1])

v1 = M @ u1
v1
```

```
array([-4.,  3.])
```

The angle of rotation is the difference between the angle of the output vector from the x-axis and the angle (in degrees) of the input vector from the x-axis (the latter of which we know is 90°):

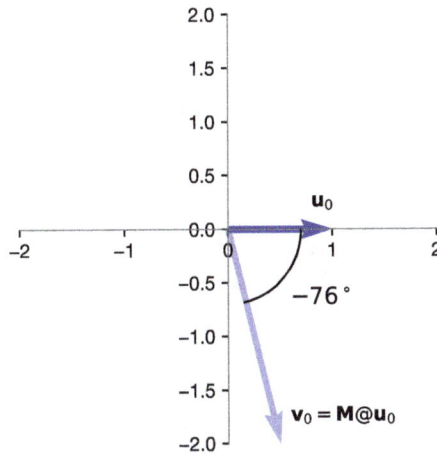

Fig. 3.2: Transformation of the unit vector $\mathbf{x} = [1, 0]^\mathsf{T}$ by the matrix $\mathbf{M}$.

```
np.rad2deg(np.arctan2(v1[1], v1[0])) - np.rad2deg(np.arctan2(u1[1], u1[0]))
```

```
53.13010235415598
```

The input and output vectors in this case are shown in Fig. 3.3. We see that not only are the vectors rotated by different amounts– they are even rotated in different directions! In addition, the amount of scaling is different. Again, the input vector length is 1, while the output vector's length is

```
np.linalg.norm(v1)
```

```
5.0
```

Exercise

Let $\mathbf{w}$ be a unit vector that is at angle ϕ from the positive x-axis. Try to modify the angle phi in the following code to:

1. Find a value of phi such that the output vector and the input vector are at the same angle. Make note of that value of phi and the length of the output vector.

2. Find a value of phi such that the output vector is in the opposite direction of the input vector (i.e., the angle between the vectors is ±180°). Make note of that value of phi and the length of the output vector.

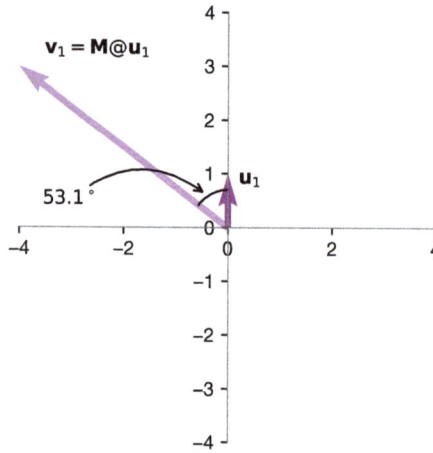

Fig. 3.3: Transformation of the unit vector $\mathbf{y} = [1,0]^{\mathsf{T}}$ by the matrix $\mathbf{M}$.

```python
# input vector angle (in degrees)
phi = 90

# create the input vector
phi_rad = np.deg2rad(phi)
w = np.array( [ np.cos(phi_rad), np.sin(phi_rad) ] )

# find output vector
z = M @ w

# find rotation from linear tranform and length of output vector
print('Rotation between input and output:',
      f'{np.rad2deg(np.arctan2(z[1], z[0]))  - np.rad2deg(np.arctan2(w[1], w[0]))}:
.2f}')
print(f'Length of output vector: {np.linalg.norm(z): .2f}')
```

```
Rotation between input and output:   53.13
Length of output vector:   5.00
```

You should have been able to find values of `phi` that satisfy each of these requirements. (In fact, there are two values in $(-180°, 180°]$ that will satisfy each of these requirements because if `phi` satisfies the requirement, so does `phi` $\pm 180°$). The unit vectors you found are called *eigenvectors* of the matrix $\mathbf{M}$. We study eigenvectors in more detail in Section 3.6.

Now consider how a linear transformation affects a region of space. To visualize this, we will represent a set of vectors within a region by points located at the head of those vectors (as we did in Section 2.3). We will use the function `transform_field()` from the PlotVec library to show the points before and after the linear transformation (multiplication by M). The `transform_field()` function generates a field of points in a square region (by default, the square of side 6 that is centered at the origin). This is shown in the plot on the left. The plot on the right shows how those points are transformed when they are treated as vectors and left-multiplied by **M**.

```
from plotvec import transform_field
transform_field(M, preserve_axes=False)
```

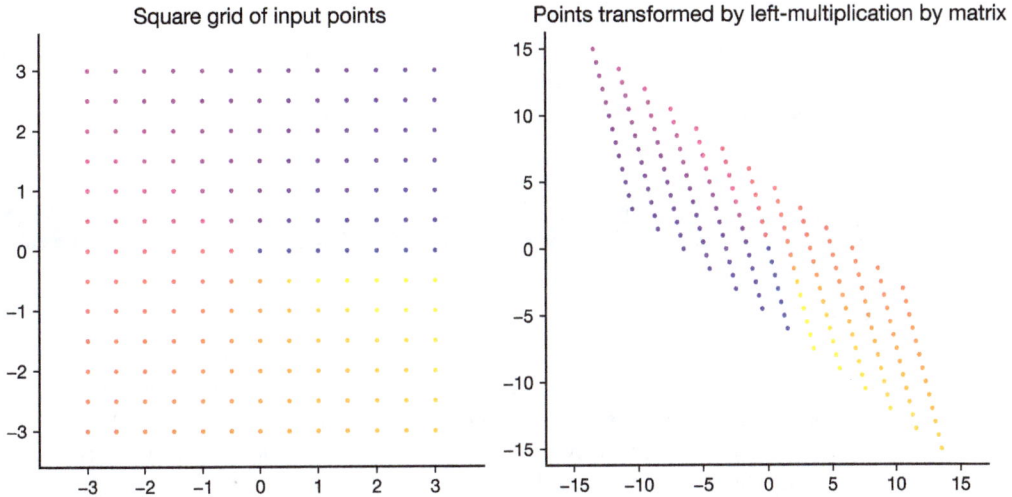

Because points in a given direction are rotated, flipped, and scaled by the same amount, the result is that the points in the square have been stretched out in space (notice the different ranges of the axes) to form a parallelogram.

In 2-D space, a linear transformation will always map the points in a square region into a parallelogram, except for some degenerate cases in which the parallelogram reduces to a line. Because the scaling factors are identical regardless of the distance of the input points from the origin, the ratio of the area of the output parallelogram to the input square is a fixed constant that depends on the matrix **M**. The ratio of these areas is the absolute value of the *determinant* of **M**. The determinant and its computation are discussed more in Section 3.5. In NumPy, the determinant function is part of the linear algebra library and can be called as `np.linalg.det()`:

```
print(f'{np.abs( np.linalg.det(M) ):.2f}')
```

```
6.50
```

Since the area of the input rectangle is $6 \times 6 = 36$, the area of the parallelogram is $36 \times 6.5 = 234$.

Terminology review and self-assessment questions

Interactive flashcards to review the terminology introduced in this section and self-assessment questions are available at la4ds.net/3-3, which can also be accessed using this QR code:

3.4 Matrix Multiplication

Linear transformations are a powerful tool in feature extraction, as we saw with the Iris data set. However, in most applications, we do not wish to project the data onto only one vector, which creates only one feature. We may want to extract multiple features for a high-dimensional data set by projecting it onto many different vectors. We could iterate over the set of vectors and compute the matrix-vector product for each one, but fortunately we do not have to do that. Matrix-matrix multiplication (we will just call this *matrix multiplication*) can be used to perform all of the products in a single operation.

Matrix multiplication can be defined as an extension of matrix-vector multiplication as follows:

DEFINITION

multiplication (matrix),
matrix product

> The *product* of matrices $\mathbf{A}$ and $\mathbf{B}$ is written $\mathbf{AB}$ and is defined if the inner dimensions of these matrices agree; that is, if $\mathbf{A}$ is a $k \times \ell$ matrix and $\mathbf{B}$ is an $\ell \times m$ matrix. In that case, the output is a $k \times m$ matrix $\mathbf{C} = \mathbf{AB}$, where the k, mth entry of $\mathbf{C}$ is given by $c_{k,m} = \mathbf{a}_{k*} \cdot \mathbf{b}_m$ (the dot product of the kth row of $\mathbf{A}$ with the mth column of $\mathbf{B}$).

To understand the definition of matrix multiplication better, let's see how to compute a matrix product by hand.

3.4.1 Computation of Matrix Product by Hand

The algorithm for computing a matrix product by hand is very similar to the algorithm for computing a matrix-vector product. Just as in matrix-vector multiplication, we iterate down the rows. However, in matrix-vector multiplication, we compute one dot product for each row. In matrix multiplication, for each row in the left-hand matrix, we compute dot products with each of the columns in the right-hand matrix. Let's extend our previous example to multiply our 3×2 matrix $\mathbf{M}$ by a 2×2 matrix $\mathbf{N}$, where

$$\mathbf{N} = \begin{bmatrix} 2 & 1 \\ -1 & -2 \end{bmatrix}.$$

The product, which we will denote as $\mathbf{W}$, has the outer dimensions of the two matrices, $\boxed{3} \times 2$ by $2 \times \boxed{2}$; thus, $\mathbf{W}$ is a 3×2 matrix. We can write the matrix product as

$$\begin{bmatrix} 3 & 4 \\ -1 & 2 \\ 2 & 3 \end{bmatrix} \begin{bmatrix} 2 & 1 \\ -1 & -2 \end{bmatrix} = \begin{bmatrix} w_{0,0} & w_{0,1} \\ w_{1,0} & w_{1,1} \\ w_{2,0} & w_{2,1} \end{bmatrix}.$$

As before, we start with the first row (row 0) of the left matrix in the matrix product. We begin by computing the output-matrix element $w_{0,0}$ by computing the dot product of row 0 of the left matrix with column 0 of the right matrix. Since column 0 is the same as the vector used in our previous matrix-vector example, the result is the same:

$$\begin{bmatrix} 3 & 4 \\ -1 & 2 \\ 2 & 3 \end{bmatrix} \begin{bmatrix} 2 & 1 \\ -1 & -2 \end{bmatrix} = \begin{bmatrix} w_{0,0} & w_{0,1} \\ w_{1,0} & w_{1,1} \\ w_{2,0} & w_{2,1} \end{bmatrix}$$

$$w_{0,0} = 3(2) + 4(-1) = 2$$

To compute the element $w_{0,1}$, we compute the dot product of row 0 of the left matrix with column 1 of the right matrix:

$$\begin{bmatrix} 3 & 4 \\ -1 & 2 \\ 2 & 3 \end{bmatrix} \begin{bmatrix} 2 & 1 \\ -1 & -2 \end{bmatrix} = \begin{bmatrix} 2 & w_{0,1} \\ w_{1,0} & w_{1,1} \\ w_{2,0} & w_{2,1} \end{bmatrix}$$

$$w_{0,1} = 3(1) + 4(-2) = -5$$

The components $w_{1,0}$ and $w_{2,0}$ will be the same as the dot products of rows 1 and 2, respectively, of $\mathbf{M}$ with the vector $\mathbf{y}$, so in each of the following visual representations, we include both those computations as well as the corresponding element $w_{1,1}$ or $w_{2,1}$. Starting with row 1 of $\mathbf{M}$, we compute the dot products of that row with each of the columns of $\mathbf{U}$ to get $w_{1,0}$ and $w_{1,1}$ as shown:

$$\begin{bmatrix} 3 & 4 \\ -1 & 2 \\ 2 & 3 \end{bmatrix} \begin{bmatrix} 2 & 1 \\ -1 & -2 \end{bmatrix} = \begin{bmatrix} 2 & -5 \\ w_{1,0} & w_{1,1} \\ w_{2,0} & w_{2,1} \end{bmatrix}$$

$$w_{1,0} = -1(2) + 2(-1) = -4$$
$$w_{1,1} = -1(1) + 2(-2) = -5$$

Finally, row 2 is used to compute $w_{2,0}$ and $w_{2,1}$:

$$\begin{bmatrix} 3 & 4 \\ -1 & 2 \\ 2 & 3 \end{bmatrix} \begin{bmatrix} 2 & 1 \\ -1 & -2 \end{bmatrix} = \begin{bmatrix} 2 & -5 \\ -4 & -5 \\ w_{2,0} & w_{2,1} \end{bmatrix}$$

$$w_{2,0} = 2(2) + 3(-1) = 1$$
$$w_{2,1} = 2(1) + 3(-2) = -4$$

The resulting matrix product values are:

$$\begin{bmatrix} 3 & 4 \\ -1 & 2 \\ 2 & 3 \end{bmatrix} \begin{bmatrix} 2 & 1 \\ -1 & -2 \end{bmatrix} = \begin{bmatrix} 2 & -5 \\ -4 & -5 \\ 1 & -4 \end{bmatrix}$$

Matrix multiplication in Python also uses the @ sign, and we can check our work easily using NumPy:

```python
M = np.array([[3, 4],
              [-1, 2],
              [2, 3]])
V = np.array([[2, 1],
              [-1, -2]])
print(M @ V)
```

```
[[ 2 -5]
 [-4 -5]
 [ 1 -4]]
```

Knowing how to perform matrix multiplication by hand is a useful skill for engineers and scientists and will be helpful to data scientists learning how machine learning algorithms work. However, in most cases, a computer or calculator should be used for computing such products to avoid errors in carrying out the many computations.

3.4.2   Properties of Matrix Multiplication

Note from the above discussion that the i, jth output element is always the dot product of the ith row of the left matrix with the jth column of the right matrix. Then consider arbitrary matrices $\mathbf{A}$ and $\mathbf{B}$, where $\mathbf{A}$ has dimensions $k \times \ell$ and $\mathbf{B}$ has dimensions $\ell \times m$, the output is the $k \times m$ matrix with the following form:

$$\begin{bmatrix} a_{0*} \cdot b_0 & a_{0*} \cdot b_1 & \dots & a_{0*} \cdot b_{m-1} \\ a_{1*} \cdot b_0 & a_{1*} \cdot b_1 & \dots & a_{1*} \cdot b_{m-1} \\ \vdots & \vdots & & \vdots \\ a_{k-1*} \cdot b_0 & a_{k-1*} \cdot b_1 & \dots & a_{k-1*} \cdot b_{m-1} \end{bmatrix}.$$

> **!**
>
> ### Important!
>
> **Order Matters!**
> Note that order is very important in matrix multiplication. In general, $\mathbf{AB} \neq \mathbf{BA}$. In fact, in many cases, one of these products may be defined while the other is not defined because matrix multiplication requires that the inner dimensions of the matrices agree.

For instance, consider the matrices $\mathbf{M}$ and $\mathbf{V}$ in the example above. The matrix product $\mathbf{VM}$ is not defined because $\mathbf{V}$ is $2 \times \boxed{2}$ and $\mathbf{M}$ is $\boxed{3} \times 2$, so the inner dimensions (shown boxed) do not agree. If we try to perform this multiplication in NumPy, it throws an error:

```
V @ M
```

```
------------------------------------------------------------
ValueError                          Traceback (most recent call last)
Cell In[6], line 1
----> 1 V @ M

ValueError: matmul: Input operand 1 has a mismatch in its core dimension 0,
with gufunc signature (n?,k),(k,m?)->(n?,m?) (size 3 is different from 2)
```

Even if the dimensions allow the order to be swapped, the product $\mathbf{AB}$ is generally not equal to $\mathbf{BA}$. For instance, let's append another column to $\mathbf{V}$ to create a 2×3 matrix, $\mathbf{Q}$:

```
new_col = np.array([[-4, 3]]).T
Q = np.hstack( (V, new_col))
print(Q)
```

```
[[ 2  1 -4]
 [-1 -2  3]]
```

Then not only is $\mathbf{MQ} \neq \mathbf{QM}$, but the dimensions of the product even differ:

- The ordered dimensions for $\mathbf{MQ}$, with the outer dimensions boxed, are $\boxed{3} \times 2$ and $2 \times \boxed{3}$. The product $\mathbf{MQ}$ will have dimension 3×3.

- The ordered dimensions for $\mathbf{QM}$, with the outer dimensions boxed, are $\boxed{2} \times 3$ and $3 \times \boxed{2}$. The product $\mathbf{MQ}$ will have dimension 2×2.

Let's use NumPy to compute these products:

```
print("MQ = ")
print(M @ Q)
print()

print("QM = ")
print(Q @ M)
```

```
MQ =
[[ 2 -5  0]
 [-4 -5 10]
 [ 1 -4  1]]

QM =
[[-3 -2]
 [ 5  1]]
```

Multiplication with the Identity Matrix

If we left- or right-multiply any matrix $\mathbf{A}$ by an appropriately-sized identity matrix, the result is the matrix $\mathbf{A}$. Examples are shown below using NumPy:

```
print('M = ')
print(M, '\n')

print('MI = ')
print( M @ np.eye(2, dtype=int), '\n' )

print("IM = ")
print( np.eye(3, dtype=int) @ M  )
```

```
M =
[[ 3  4]
 [-1  2]
 [ 2  3]]

MI =
[[ 3  4]
 [-1  2]
 [ 2  3]]

IM =
[[ 3  4]
 [-1  2]
 [ 2  3]]
```

Multiplication and Transpose

Recall again the form of the product $\mathbf{AB}$, where $\mathbf{A}$ has dimensions $k \times \ell$ and $\mathbf{B}$ has dimensions $\ell \times m$,

$$\begin{bmatrix} a_{0*} \cdot b_0 & a_{0*} \cdot b_1 & \dots & a_{0*} \cdot b_{m-1} \\ a_{1*} \cdot b_0 & a_{1*} \cdot b_1 & \dots & a_{1*} \cdot b_{m-1} \\ \vdots & \vdots & & \vdots \\ a_{k-1*} \cdot b_0 & a_{k-1*} \cdot b_1 & \dots & a_{k-1*} \cdot b_{m-1} \end{bmatrix}.$$

If we take transposes of $\mathbf{A}$ and $\mathbf{B}$, then the matrix product is not generally defined because $\mathbf{A}^\mathsf{T}$ is $\ell \times k$ and $\mathbf{B}^\mathsf{T}$ is $m \times \ell$. Moreover, the product $\mathbf{A}^\mathsf{T}\mathbf{B}^\mathsf{T}$ would now be equivalent to multiplying the *columns* of $\mathbf{A}$ with the *rows* of $\mathbf{B}$, which would be very different than multiplying the rows of $\mathbf{A}$ with the columns of $\mathbf{B}$.

However, consider the product $\mathbf{B}^\mathsf{T}\mathbf{A}^\mathsf{T}$. The ordered dimensions agree: $m \times \ell$ and $\ell \times k$, and the dimensions of the product will be $m \times k$. The rows of $\mathbf{B}^\mathsf{T}$ are the columns of $\mathbf{B}$, and the columns of $\mathbf{A}^\mathsf{T}$ are the rows of $\mathbf{A}$. Dot product commutes (does not care about order), so each product of a row of $\mathbf{B}^\mathsf{T}$ and a column of $\mathbf{A}^\mathsf{T}$ is one of the components in $\mathbf{AB}$. We can write the product $\mathbf{B}^\mathsf{T}\mathbf{A}^\mathsf{T}$ as dot products of the rows of $\mathbf{A}$ and columns of $\mathbf{B}$ as

$$\begin{aligned} \mathbf{B}^\mathsf{T}\mathbf{A}^\mathsf{T} &= \begin{bmatrix} b_0 \cdot a_{0*} & b_0 \cdot a_{1*} & \dots & b_0 \cdot a_{k-1*} \\ b_1 \cdot a_{0*} & b_1 \cdot a_{1*} & \dots & b_1 \cdot a_{k-1*} \\ \vdots & \vdots & & \vdots \\ b_{m-1} \cdot a_{0*} & b_{m-1} \cdot a_{1*} & \dots & b_{m-1} \cdot a_{k-1*} \end{bmatrix} \\ &= \begin{bmatrix} a_{0*} \cdot b_0 & a_{1*} \cdot b_0 & \dots & a_{k-1*} \cdot b_0 \\ a_{0*} \cdot b_1 & a_{1*} \cdot b_1 & \dots & a_{k-1*} \cdot b_1 \\ \vdots & \vdots & & \vdots \\ a_{0*} \cdot b_{m-1} & a_{1*} \cdot b_{m-1} & \dots & a_{k-1*} \cdot b_{m-1} \end{bmatrix} \\ &= (\mathbf{AB})^\mathsf{T}. \end{aligned}$$

Thus, the transpose of a product of matrices is the product of the transposes of those individual matrices in reverse order.

3.4.3 Application of Matrix Multiplication to Rotating Data

Recall the 40° rotation vector we used for feature extraction in Section 3.2.4, the vector $\tilde{\mathbf{x}} = [\cos 40°, -\sin 40°]^\mathsf{T}$. We can create a vector at 90° to this vector as $\tilde{\mathbf{y}} = [\sin 40°, \cos 40°]^\mathsf{T}$. If we stack these vectors into the columns of a matrix, then it will be an *orthogonal matrix*:

> **DEFINITION**
>
> **orthogonal matrix**
>
> A (real) square matrix whose columns are a set of orthonormal vectors.

To check if a matrix is an orthogonal matrix, we just need to confirm that the product of any two columns is equal to 0 and the product of any column with itself is equal to 1. Note that the rows of $\mathbf{U}^\mathsf{T}$ are the columns of $\mathbf{U}$. Thus, we can efficiently find all the products between all pairs of columns using matrix multiplication of the form $\mathbf{U}^\mathsf{T}\mathbf{U}$. For matrix $\mathbf{U}$ to be orthogonal, the equivalent conditions for $\mathbf{U}^\mathsf{T}\mathbf{U}$ are:

- all of the off-diagonal elements of $\mathbf{U}^\mathsf{T}\mathbf{U}$, which correspond to dot products of two different columns, should be zero, and

- all of the diagonal elements of $\mathbf{U}^\mathsf{T}\mathbf{U}$, which correspond to dot products of columns with themselves, should be one.

If $\mathbf{U}$ is an orthogonal matrix, then $\mathbf{U}^\mathsf{T}\mathbf{U} = \mathbf{I}$. Note that $\mathbf{U}^\mathsf{T}\mathbf{U}$ is equal to $\mathbf{U}\mathbf{U}^\mathsf{T}$ because the transpose of $\mathbf{I}$ is still $\mathbf{I}$. So, we can check whether a matrix is an orthogonal matrix by checking whether $\mathbf{U}\mathbf{U}^\mathsf{T} = \mathbf{I}$.

Let's check whether the matrix of rotation vectors is an orthogonal matrix:

```
cos40 = np.cos( np.deg2rad(40) )
sin40 = np.sin( np.deg2rad(40) )

x40cw = np.array([[cos40, -sin40]]).T
y40cw = np.array([[sin40, cos40]]).T

U = np.hstack( (x40cw, y40cw) )
```

Here we used `np.hstack()` to horizontally stack the column vectors into a matrix. Note that `np.hstack()`'s argument is a **tuple** containing the vectors or matrices to be horizontally stacked together.

Let's calculate $\mathbf{U}^\mathsf{T}\mathbf{U}$ and $\mathbf{U}\mathbf{U}^\mathsf{T}$ for this matrix:

```
print(np.round(U.T @ U, 10), '\n')
print(np.round(U @ U.T , 10))
```

```
[[1. 0.]
 [0. 1.]]

[[ 1. -0.]
 [-0.  1.]]
```

We see that either check confirms that **U** is an orthogonal matrix. In fact, this is a special type of orthogonal matrix called a rotation matrix. It is easy to see the effect of this rotation matrix as a linear transformation by looking at the output of `transform_field(U, preserve_axes=False)`, which is shown in Fig. 3.4. I have fixed the axis limits to be equal. The output field of points is a rotated version of the input field of points.

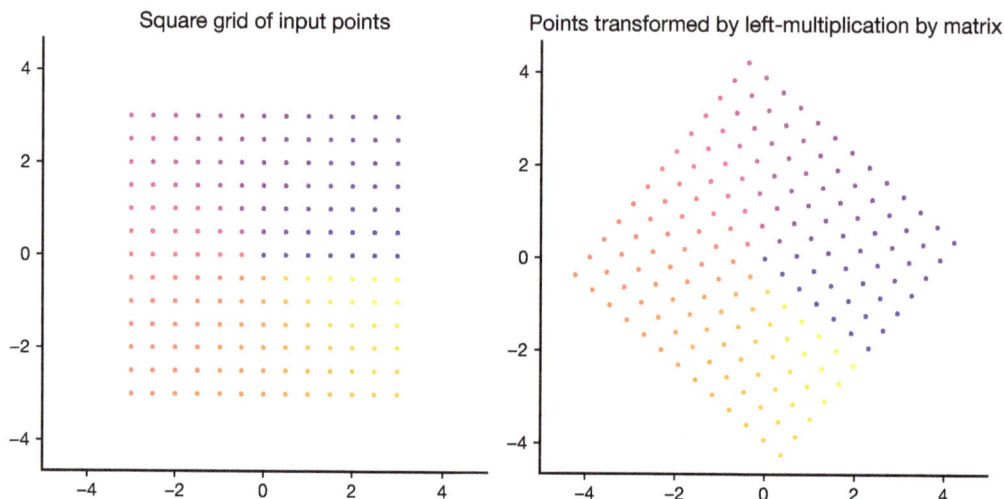

Fig. 3.4: Visualization of linear transformation of points for the matrix **U** created from rotated basis vectors.

Gram Matrix

In our test for an orthogonal matrix, we found the inner products of all pairs of columns of **U**. We can apply this same calculation to **any** matrix **A** to get the *Gram matrix*:

> **DEFINITION**
>
> **Gram matrix**
>
> For a matrix **A**, the Gram matrix is the matrix whose i, jth entry is the dot product of columns i and j of **A**. The Gram matrix can be calculated as $\mathbf{A}^\mathsf{T}\mathbf{A}$.

Let **A** be a $k \times m$ matrix, and let a_i denote the ith column of **A**. Then the Gram matrix has the following form:

$$\mathbf{A}^\mathsf{T}\mathbf{A} = \begin{bmatrix} a_0 \cdot a_0 & a_0 \cdot a_1 & \cdots & a_0 \cdot a_{m-1} \\ a_1 \cdot a_0 & a_1 \cdot a_1 & \cdots & a_1 \cdot a_{m-1} \\ \vdots & \vdots & & \vdots \\ a_{m-1} \cdot a_0 & a_{m-1} \cdot a_1 & \cdots & a_{m-1} \cdot a_{m-1} \end{bmatrix}.$$

Since dot product is commutative, we see that the i, jth element and the j, ith element of the Gram matrix are equal. The Gram matrix is a *symmetric matrix*:

> **DEFINITION**
>
> **symmetric matrix**
> A matrix $\mathbf{M}$ is symmetric if the i, jth element is equal to the j, ith element for any valid i and j. Equivalently, $\mathbf{M}^\mathsf{T} = \mathbf{M}$.

Let's compute the Gram matrix for the matrix $\mathbf{M}$ that we have been using in our examples:

```
print(M.T @ M)
```

```
[[14 16]
 [16 29]]
```

We see that the Gram matrix is symmetric, as expected. Note that the Gram matrix of $\mathbf{N} = \mathbf{M}^\mathsf{T}$ is not the same as the Gram matrix of $\mathbf{M}$ because it is equivalent to finding the inner products of all of the **rows** of $\mathbf{M}$.

```
N = M.T
print(N.T @ N)
```

```
[[25  5 18]
 [ 5  5  4]
 [18  4 13]]
```

Example Using Real Data: Extracting Multiple Features

Let's see how to use matrix multiplication to extract multiple features simultaneously. For simplicity, I will again use the Iris data but now create two features by projecting the data onto each of the orthogonal vectors $\tilde{\mathbf{x}}$ and $\tilde{\mathbf{y}}$ from the example above.

If we want to get both features for the rotated data, we can use matrix-vector multiplication twice, as shown in the code below:

```
from sklearn import datasets
iris = datasets.load_iris()
class01 = np.where(iris['target']<2)[0]
R = iris.data[class01][:,:2]

new_feature0 = R @ x40cw
new_feature1 = R @ y40cw
```

However, we could instead compute both features simultaneously using matrix multiplication as

```
new_features = R @ U
print(new_features[:5])
```

```
[[1.65707001 5.95937235]
 [1.82525493 5.44779261]
 [1.54348852 5.47244398]
 [1.53116283 5.33156077]
 [1.5161868  5.97169803]]
```

If we compare with the outputs of the separate dot products, we see that the results are the same. In general, the matrix multiplication version will be faster because it can take advantage of vectorized computations in the microprocessor and has the additional benefit that the results are stored together in a single NumPy array.

```
print(new_feature0[:5])
print()
print(new_feature1[:5])
```

```
[[1.65707001]
 [1.82525493]
 [1.54348852]
 [1.53116283]
 [1.5161868 ]]

[[5.95937235]
 [5.44779261]
 [5.47244398]
 [5.33156077]
 [5.97169803]]
```

Terminology review and self-assessment questions

Interactive flashcards to review the terminology introduced in this section and self-assessment questions are available at la4ds.net/3-4, which can also be accessed using this QR code:

3.5   Matrix Determinant and Linear Transformations

In Section 3.2.4, we found that an $n \times n$ square matrix can be seen as a linear transformation from $\mathbb{R}^n$ to $\mathbb{R}^n$. I showed by example that such a linear transformation stretches or compresses space, and I claimed that the amount of that stretch is related to a property of square matrices called the *determinant*:

> **DEFINITION**
>
> **determinant**
>
> For a square matrix $\mathbf{M}$, the determinant is a scalar value that is related to how $\mathbf{M}$ stretches or compresses space if it is used as a linear transformation. The determinant of $\mathbf{M}$ is denoted by $\det \mathbf{M}$ or $|\mathbf{M}|$. The determinant may be positive, negative, or zero.

Note that I did not include a formula for the determinant. That is because there is no simple formula for the determinant of general $n \times n$ matrices. In this section, I will only teach you how to find the determinant for 2×2 matrices by hand. There is also a simple approach for finding the determinant of 3×3 matrices, but in most cases, determinants should be found using computers or calculators. In a later section, I will show you how to find the determinant for any matrix by finding a related matrix that is in upper triangular form.

Finding the determinant for a 2×2 matrix is easy: multiply the diagonal elements and subtract the product of the off-diagonal elements, as shown below:

$$\det \mathbf{M} = \det \begin{bmatrix} a & b \\ c & d \end{bmatrix} = \begin{vmatrix} a & b \\ c & d \end{vmatrix}$$

$$= ad - bc.$$

We can use the NumPy function `np.linalg.det()` or the PyTorch function `torch.linalg.det()` to calculate the determinant of arrays or tensors, respectively. If we have a SymPy Matrix object, we can call the `det()` method. I give examples for NumPy and SymPy below.

Below I show several examples of 2×2 matrices and show how to calculate the determinant for each. Each determinant is checked using NumPy. Then I use a plot to show how each matrix translates a uniformly spaced set of points in a rectangle of 2-D space to another 2-D space. Finally, I discuss the interpretation of the determinant with respect to how the matrix transforms points in 2-D space.

For each of the matrices, I plot the location of the points shown in Fig. 3.5 after the linear transformation. We again use the `plot_field()` function from the PlotVec library that takes a square field of points and plots the points after a linear transformation. Fig. 3.5 shows the default field of points at the input, which is the output of `plot_field()` with the default `matrix` parameter, which is $\mathbf{I}_2$, a 2×2 identity matrix. The code to generate all of the plots of fields of points is available online at la4ds.net/3-5. Note that the area of the input field of points is $[3 - (-3)]\,[(3 - (-3)] = 36$.

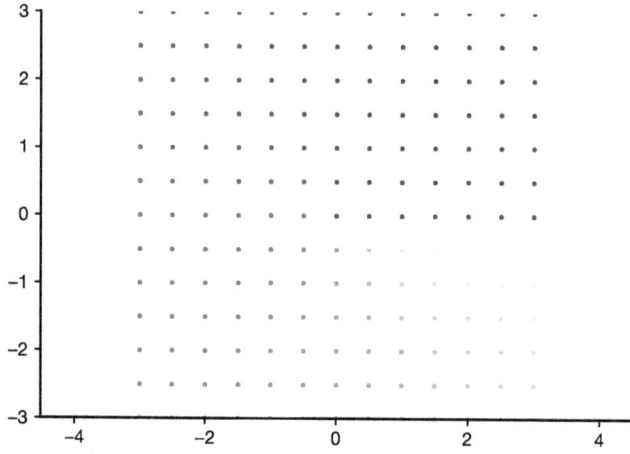

Fig. 3.5: Field of points uniformly spaced over the region $[-3,3] \times [-3,3]$.

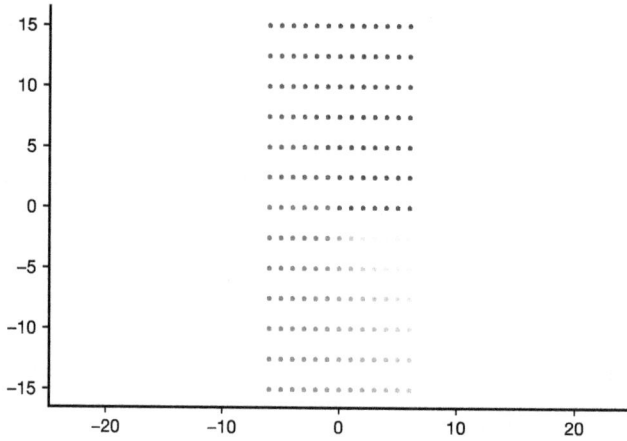

Fig. 3.6: Field of points after transformation by diagonal matrix with components greater than 1.

Example 3.4: Multiplication by a Diagonal Matrix with Components Greater than 1

Let's start with a simple case: a diagonal matrix with positive components that are greater than 1:

$$\mathbf{M}_1 = \begin{bmatrix} 2 & 0 \\ 0 & 5 \end{bmatrix}.$$

The transformed field of points is shown in Fig. 3.6, where the range of the axes was increased to accommodate the transformed field of points. The new field is still a rectangle, with lower-left corner (-6, -15) and upper-right corner (6, 15), and no rotation of the points has occurred. The x-coordinates have been expanded by a factor of 2, and the y-coordinates have been expanded by a factor of 5, corresponding to the components on the diagonal of $\mathbf{M}_1$: a diagonal matrix individually scales the x- and y-coordinates.

The area of the region is $12 \times 30 = 360$, which is 10 times larger than the area of the original field. The determinant calculation is shown below:

$$\begin{vmatrix} 2 & 0 \\ 0 & 5 \end{vmatrix} = 2 \cdot 5 - 0 \cdot 0 = 10.$$

The determinant is equal to the factor by which the linear transformation changed the area of the field of points. Let's check using NumPy:

```
M1 = np.diag([2,5])
print(M1)
print()
print(f'|M1| = {np.linalg.det(M1): .1f}')
```

```
[[2 0]
 [0 5]]

|M1| =  10.0
```

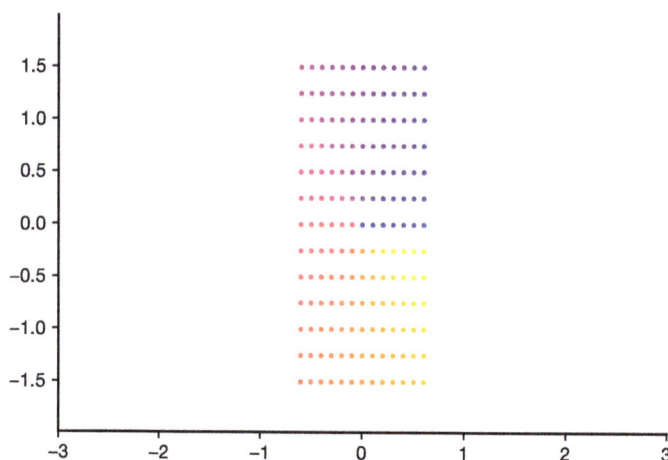

Fig. 3.7: Field of points after transformation by diagonal matrix with positive components that are smaller than 1.

Example 3.5: Multiplication by a Diagonal Matrix with Positive Components Smaller than 1

Now consider a diagonal matrix with positive components that are smaller than 1:

$$\mathbf{M}_2 = \begin{bmatrix} 0.2 & 0 \\ 0 & 0.5 \end{bmatrix}.$$

Fig. 3.7 shows the field of points after the linear transformation. The linearly transformed points fit in a small subregion of the original area. As expected from our previous example, the x- and y-components are individually scaled by the corresponding components of the diagonal matrix. Thus, the field is scaled by 0.2 in the x-dimension and 0.5 in the y-dimension. The ratio of the resulting area to the area of the original field of points is 0.1, which is the determinant of $\mathbf{M}_2$:

$$\begin{vmatrix} 0.2 & 0 \\ 0 & 0.5 \end{vmatrix} = 0.2 \cdot 0.5 - 0 \cdot 0 = 0.1.$$

Let's check using NumPy:

```
M2 = np.diag([0.2, 0.5])
print(f'|M2| = {np.linalg.det(M2):.2f}')
```

```
|M2| = 0.10
```

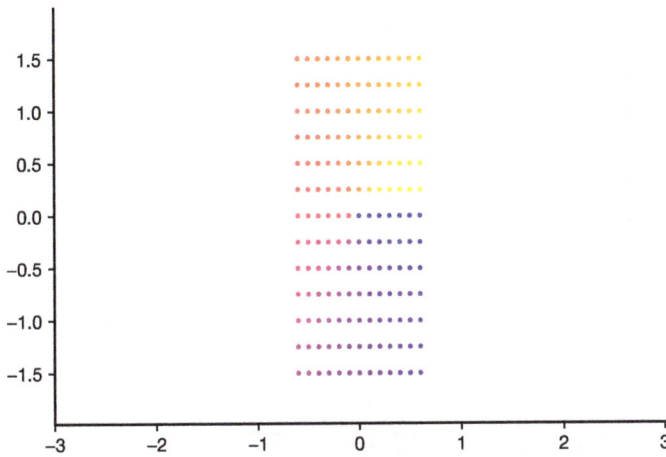

Fig. 3.8: Field of points after transformation by diagonal matrix with positive and negative components with magnitudes smaller than 1.

Example 3.6: Multiplication by a Diagonal Matrix with Positive and Negative Components

Next consider what happens if one of the elements is negative. Consider the matrix

$$\mathbf{M}_3 = \begin{bmatrix} 0.2 & 0 \\ 0 & -0.5 \end{bmatrix}.$$

The field of transformed points is shown in Fig. 3.8. If you compare the colors of the points in the original field of points with those in the transformed field, it should be obvious the field of points has been flipped in the y direction (i.e., around the x-axis). The factor of -0.5 results in each positive y-component at the input being mapped to a negative value at the output, and each negative y-component at the input being mapped to a positive value at the output.

The area is still scaled by 0.1, which is the **absolute value** of the determinant, $0.2 \cdot (-0.5) - 0 \cdot 0 = -0.1$. We can verify using NumPy:

```python
M3 = np.diag([0.2, -0.5])
print(f'|M3| = {np.linalg.det(M3):.2f}')
```

```
|M3| = -0.10
```

Example 3.7: Multiplication by an Orthogonal Matrix

Now let's consider the orthogonal matrix $\mathbf{U}$ from Section 3.4:

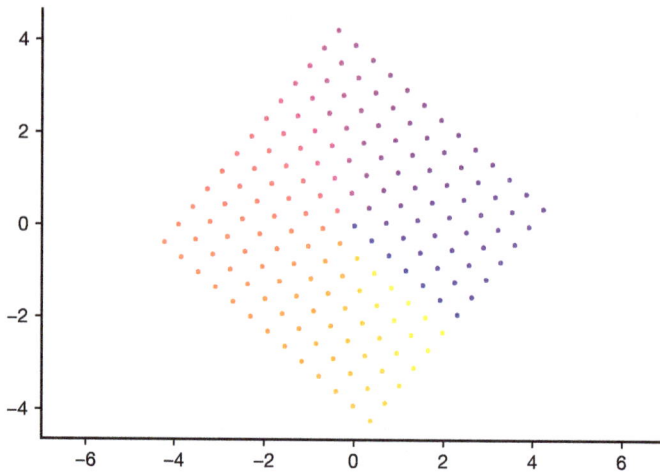

Fig. 3.9: Field of points after transformation by an orthogonal matrix.

```
U = np.array([[ 0.76604444,  0.64278761],
              [-0.64278761,  0.76604444]])
```

The transformed field of points is shown in Fig. 3.9. From the figure, it appears that the linear transformation using **U** does not scale the area of the field of points but only rotates it by 40° clockwise.

If the area of the field of points is unchanged, the determinant should be 1. The determinant of this matrix is $(0.76604444)^2 - (0.64278761)(-0.64278761) = 1$. Let's check using NumPy:

```
print(f'Determinant of U matrix: {np.linalg.det(U):.1f}')
```

```
Determinant of U matrix: 1.0
```

This is a general property that we will prove later: **The determinant of an orthogonal matrix is 1.**

Example 3.8: General Linear Transformation

Consider again the matrix used to demonstrate a linear transformation of a field of points in Section 3.3. This linear transformation rotates, scales, and flips the points in the field, transforming the original square region into a parallelogram, as shown in Fig. 3.10. It can be shown that the area of the parallelogram is equal to the product of the absolute value of the determinant and the area of the original region, which is shown below:

$$\begin{vmatrix} 0.5 & -4 \\ -2 & 3 \end{vmatrix} = 0.5 \cdot 3 - (-4) \cdot (-2) = -6.5.$$

We can confirm this using NumPy:

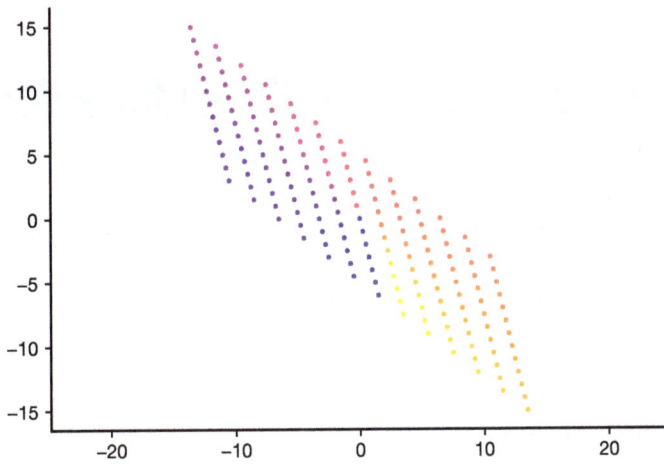

Fig. 3.10: Field of points after transformation by a general matrix.

```
M5 = np.array( [[0.5, -4],
                [ -2,  3]] )
np.linalg.det(M5)
print(f'determinant of matrix M5 = {np.linalg.det(M5):.1f}')
```

```
determinant of matrix M5 = -6.5
```

Thus the area of the parallelogram is equal to $36\,|-6.5| = 234$.

Example 3.9: Transformation by a Singular Matrix

Consider the following matrix and its determinant:

```
M6 = np.array([[1, -2],
               [2,  -4]])
print(f'determinant of matrix M6 = {np.linalg.det(M6):.1f}')
```

```
determinant of matrix M6 = 0.0
```

M6 is said to be *singular*:

DEFINITION

singular (matrix)

A square matrix **M** is *singular* if det **M** = 0.

The other matrices in our examples are nonsingular:

DEFINITION

nonsingular (matrix)

A square matrix $\mathbf{M}$ is *nonsingular* if det $\mathbf{M} \neq 0$.

Let's plot the field of transformed points to see the effect of multiplication by a singular matrix:

```
plot_field(M6, preserve_axes=False)
```

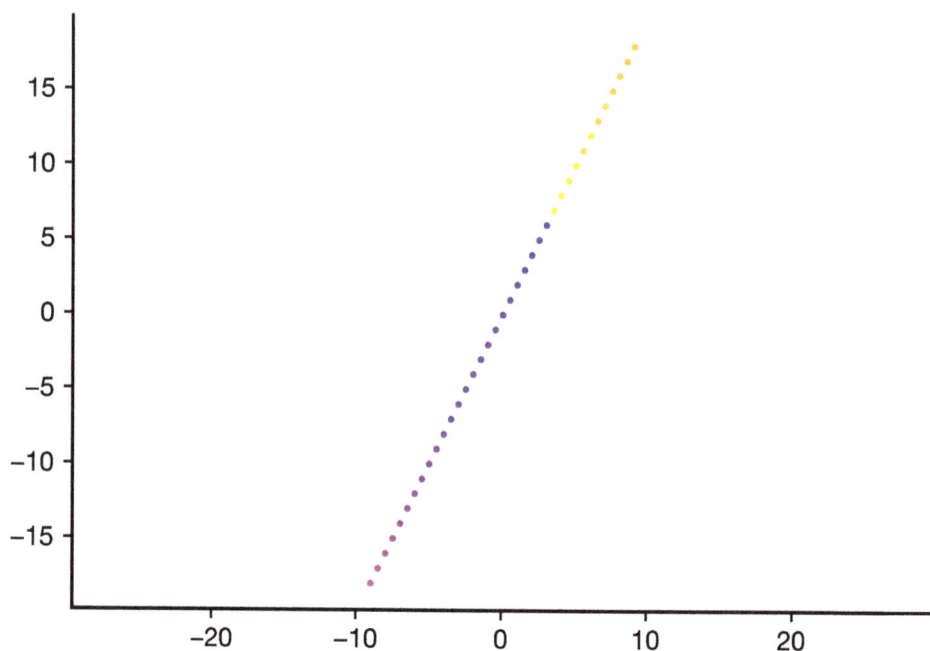

The rectangular region of points has been mapped to a line segment. The output points all lie along a single dimension within the two-dimensional output space. In general, a linear transformation that can be represented using a singular matrix will map to a lower dimensional subspace within the output space. I will provide additional interpretation of what it means for a matrix to be singular/nonsingular in Section 4.2.

3.5.1 Properties of the Determinant

The determinant has many useful properties, but here are a few of the most important ones:

1. **The determinant of an identity matrix is 1.**

2. **The determinant of a matrix and the determinant of the transpose of a matrix are the same:** $\det M^{\mathsf{T}} = \det M$.

3. **The determinant of a matrix product is the product of the determinants:** $\det(\mathbf{AB}) = (\det \mathbf{A})(\det \mathbf{B})$.

4. **For a triangular matrix, the determinant is the product of the elements on the main diagonal.** This property is useful for finding the determinant without any special formula because we will show how to put any matrix in upper-triangular form in Section 4.1.

Terminology review and self-assessment questions

Interactive flashcards to review the terminology introduced in this section and self-assessment questions are available at la4ds.net/3-5, which can also be accessed using this QR code:

3.6 Eigenvalues and Eigenvectors

Consider a square $n \times n$ matrix $\mathbf{M}$ that can be interpreted as a linear transformation from $\mathbb{R}^n \to \mathbb{R}^n$. As shown in Section 3.3, a matrix $\mathbf{M}$ can transform different vectors in different ways. In particular, we can see from the output of the `transform_unit_vecs()` command in Section 3.3 that for a two-dimensional space, vectors at different angles are scaled and rotated by different amounts.

For a nonsingular matrix, we can find output-vectors at every angle. Since we can create input vectors at every angle, there may be some input vectors for which the output vectors are at the same angle as the input vectors. If $\mathbf{u}$ is such a vector, then $\mathbf{Mu}$ should be in the same direction as $\mathbf{u}$ but may have a different length. Equivalently, $\mathbf{Mu}$ should be a scaled version of $\mathbf{u}$. Mathematically, such a vector would satisfy $\mathbf{Mu} = \lambda\mathbf{u}$ for some real constant λ. We call such a vector an *eigenvector* of $\mathbf{M}$:

DEFINITION

eigenvector

Given a square $n \times n$ matrix $\mathbf{M}$, a non-zero n-vector $\mathbf{u}$ is an *eigenvector* of $\mathbf{M}$ if there exists a real constant λ such that

$$\mathbf{Mu} = \lambda\mathbf{u} \tag{3.2}$$

for some real constant λ.

Here, *eigen* is the German word for *own*, meaning that for a given matrix, such vectors are special characteristics of that matrix. Any value λ that satisfies this equation for some eigenvector $\mathbf{u}$ is called an *eigenvalue* of $\mathbf{M}$:

> **DEFINITION**
>
> **eigenvalue**
>
> Given a square $n \times n$ matrix $\mathbf{M}$, a constant λ is an *eigenvalue* of $\mathbf{M}$ if there exists a non-zero vector $\mathbf{u}$ such that
>
> $$\mathbf{M}\mathbf{u} = \lambda\mathbf{u}.$$

> **Important!**
>
> There may be multiple vectors $\mathbf{u}$ that satisfy (3.2), so we add a subscript i to distinguish them. Each eigenvector $\mathbf{u}_i$ has an associated eigenvalue λ_i to satisfy (3.2). It is best to think of them as an eigenvector-eigenvalue pair, $(\mathbf{u}_i, \lambda_i)$.

Suppose that $(\mathbf{u}_i, \lambda_i)$ are an eigenvector-eigenvalue pair of a matrix $\mathbf{M}$. Consider a vector $c\mathbf{u}_i$, where c is a constant. Then

$$\mathbf{M}\left(c\mathbf{u}_i\right) = c\mathbf{M}\mathbf{u}_i$$
$$= c\lambda_i\mathbf{u}_i$$
$$= \lambda_i\left(c\mathbf{u}_i\right).$$

Thus, $c\mathbf{u}_i$ is also an eigenvector of $\mathbf{M}$ with the same eigenvalue λ_i. From a geometric perspective, this makes sense because $c\mathbf{u}_i$ is in the same (or exact opposite) direction as $\mathbf{u}_i$, and so we expect a linear transformation to affect it in the same way. Because of this, when we calculate and report eigenvectors, we typically report the eigenvectors that have unit norm; we will call these the *unit eigenvectors*. But you should keep in mind that any scaled version of a unit eigenvector of a matrix is also an eigenvector of that matrix. If $(\mathbf{u}_i, \lambda_i)$ are a unit eigenvector and eigenvalue pair, then $-\mathbf{u}_i$ is also a unit eigenvector with eigenvalue λ_i, and so we only report one of $\mathbf{u}_i$ and $-\mathbf{u}_i$ when reporting unit eigenvectors. The choice of which one is reported depends on the implementation for finding the eigenvectors.

NumPy has two commands to find the (unit) eigenvector-eigenvalue pairs of a matrix. Both commands are part of NumPy's `linalg` module[1]. For convenience, we will import this module as `la`. The most general method works for arbitrary square matrices and can be called as `la.eig()`. For real symmetric matrices (or more generally complex matrices that are *Hermitian* – i.e., have complex-conjugate symmetry), the `la.eigh()` function is faster and more accurate. Both `la.eig()` and `la.eigh()` return an object that has two components:

1. The first output is a vector of the eigenvalues. We will use $\boldsymbol{\lambda}$ to denote the vector of eigenvalues and $\boldsymbol{\Lambda}$ to denote a diagonal matrix whose diagonal elements are the eigenvalues. However, in Python, `lambda` is reserved keyword, and so I will use `lam` for the eigenvalue vector.

2. The second output is a matrix with unit eigenvectors in its columns. It is called the *modal (pronounced moh-dull) matrix*, and is usually denoted by $\mathbf{U}$.

---

[1] PyTorch has equivalent commands that are part of PyTorch's `linalg` module.

> DEFINITION
>
> **modal matrix**
>
> For a square matrix **M**, the modal matrix is a matrix whose columns are the unit eigenvectors of **M**.

The outputs are aligned in the sense that the ith entry of the eigenvalue vector corresponds to the ith column of the modal matrix.

Let's practice using NumPy to find the eigenvalues and eigenvectors of a matrix with an example:

Example 3.10: Eigenvectors and Eigenvalues of a 2×2 Matrix

In Section 3.3, a matrix $\mathbf{M}_5$ was used to study the effect of a linear transformation. That section included an exercise in which you were asked to experimentally find the orientations of unit vectors that produced output vectors at the same angles. The vectors at these orientations are the eigenvectors of that matrix. The code below shows how to check your results by finding the eigenvectors using `la.eig()`:

```
import numpy.linalg as la

M5 = np.array([ [0.5, -4],
                [-2,   3] ])
lam5, U5 = la.eig(M5)
lam5, U5
```

```
(array([-1.34232922,  4.84232922]),
 array([[-0.90828954,  0.67752031],
        [-0.41834209, -0.73550406]]))
```

Let's confirm that the columns of the returned modal matrix (stored in the variable U5) are eigenvectors of $\mathbf{M}_5$. Let $\mathbf{u}_0$ and $\mathbf{u}_1$ denote the columns of $\mathbf{U}_5$. Then we first compute the output vectors $\mathbf{Mu}_0$ and $\mathbf{Mu}_1$:

```
out0 = M5 @ U5[:, 0]
out0
```

```
array([1.21922359, 0.56155281])
```

```
out1 = M5 @ U5[:, 1]
out1
```

```
array([ 3.28077641, -3.56155281])
```

Now we can check to see if the output vectors are scaled versions of $\mathbf{u}_0$ and $\mathbf{u}_1$. To test this, we can perform element-wise division on the vectors and check whether the output is of the form $c\mathbf{1}$ for some constant c. Let's start with the first output vector:

```
out0 / U5[:, 0]
```

```
array([-1.34232922, -1.34232922])
```

Not only is the output a scaled version of the input, it is equal to $\lambda_0\mathbf{u}_0$, where λ_0 is the eigenvalue that corresponds to eigenvector $\mathbf{u}_0$.

Let's check for input vector $\mathbf{u}_1$:

```
out1 / U5[:, 1]
```

```
array([4.84232922, 4.84232922])
```

We can see that the output can be written as $\lambda_1\mathbf{u}_1$.

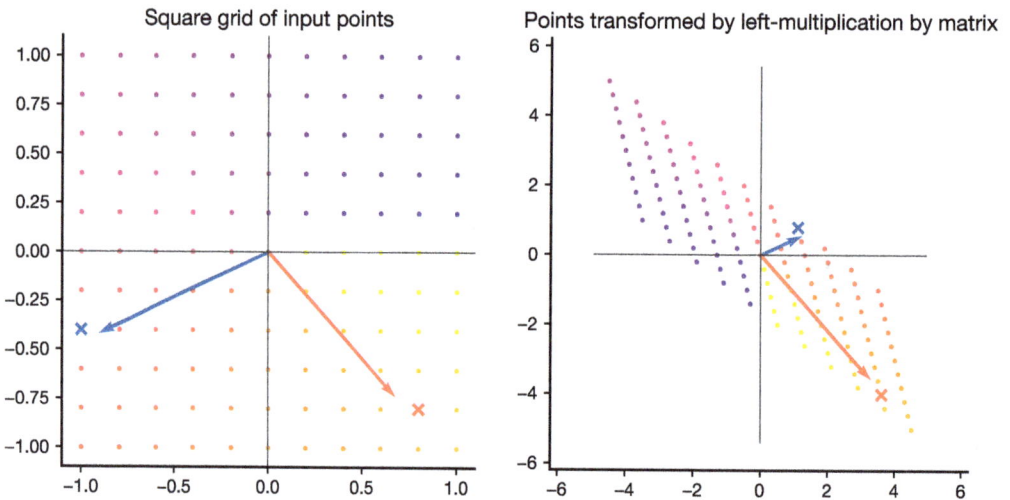

Fig. 3.11: Field of points before and after linear transformation by matrix $\mathbf{M}_5$. Input field is overlaid with unit eigenvectors; output field is overlaid with unit eigenvectors scaled by corresponding eigenvalues.

Fig. 3.11 illustrates how the eigenvectors of the matrix $\mathbf{M}_5$ relate to how that matrix linearly transform space[2]. The left subplot shows an input field of points on the rectangle with x-values and y-values from -1 to 1. Overlaid on this subplot are the unit eigenvectors for $\mathbf{M}_5$ that we found above. For each eigenvector, the closest point to the origin that is in the general direction of that eigenvector, while being beyond the head of that eigenvector, is marked with an 'x'. The right subplot shows the field of points after the linear transformation from multiplying by $\mathbf{M}_5$. The output field of points is overlaid by each eigenvector of $\mathbf{M}_5$ scaled by its corresponding eigenvalue. For each input point in the left subplot that is marked with 'x', the corresponding output point is marked with 'x' in the right subplot. It can be seen from the figure that the colors of the points in the directions of the eigenvectors are the same in each plot. The points along the eigenvector with a negative eigenvalue end up in an orientation that is flipped opposite of the origin. The negative eigenvalue indicates that the matrix $\mathbf{M}_5$ results in not just rotation and stretching, but also flipping of the points in space. The magnitudes of the eigenvalues show how much the field of points is stretched along the corresponding eigenvectors.

We can solve Equation 3.2 for the eigenvalues and eigenvectors, but to do so requires understanding how to solve systems of linear equations. We introduce the necessary techniques in Chapter 4 and show how to apply them in Section 6.4 to:

- solve for the eigenvalues and eigenvectors of a matrix,

- factor certain matrices in terms of their modal matrix and the diagonal matrix of eigenvalues,

- calculate the determinant from the eigenvalues and determine whether a matrix is singular based on its eigenvalues, and

- use the eigenvectors to represent data in a way that allows us to identify the most important information conveyed by that data.

Terminology review and self-assessment questions

Interactive flashcards to review the terminology introduced in this section and self-assessment questions are available at la4ds.net/3-6, which can also be accessed using this QR code:

3.7   Chapter Summary

This chapter introduced matrices, special types of matrices, and mathematical operations involving matrices. In particular, I showed that matrix-vector and matrix-matrix multiplication generalize and extend the dot product. I showed how matrix-vector multiplication can be used for feature extraction. I also introduced the concept of matrix multiplication as a linear transformation between vector spaces. I introduced the determinant of a matrix, which quantifies how much a matrix stretches or compresses space when used as a linear transform, and we considered multiple examples to understand these concepts in more detail. Finally, I introduced the concepts of eigenvectors and eigenvalues of a matrix, showed how to find them using NumPy, and showed how to interpret them.

Access a list of key take-aways for this chapter, along with interactive flashcards and quizzes at la4ds.net/3-7, which can also be accessed using this QR code:

4

Solving Systems of Linear Equations

Linear equations are often used to represent the relationships among different variables or features. When we have multiple linear equations describing such relationships, the set of equations is called a *system of linear equations*. In this chapter, I show how to use matrix techniques to solve systems of linear equations. I also introduce matrix inverses and their properties.

4.1 Working with Systems of Linear Equations Using Matrices and Vectors – Part 1

Matrices are very useful for working with systems of linear equations. They allow us to write such equations concisely and to solve them efficiently. Let's use some examples to motivate our work.

Let's start with this simple equation:

$$y = 4x - 2.$$

This is the equation for the line shown in Fig. 4.1.

When we plot a line to represent that equation, what we are really illustrating is the *solution set* for that equation:

> **DEFINITION**
>
> **solution set (equation/system of equations)**
>
> Given an equation, the *solution set* is the set of all points that satisfy the equation. For a system of equations, the solution set is the set of all points that simultaneously satisfy all of the equations.

The equation of this line is a *linear equation*:

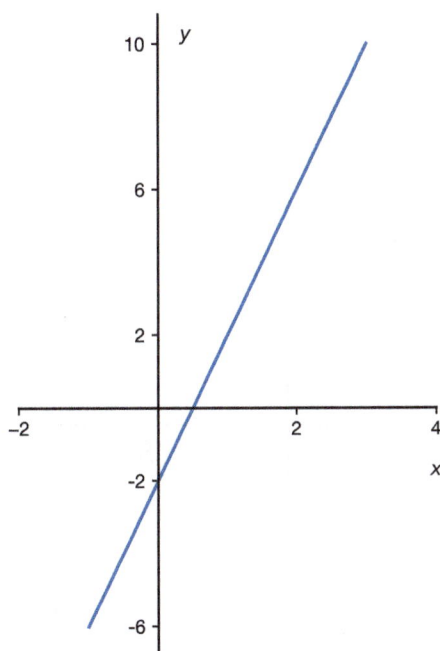

Fig. 4.1: The line given by $y = 4x - 2$.

DEFINITION

linear equation

A polynomial in one or more variables, in which all the variables have degree 1. If the variables are $x_0, x_1, \ldots, x_{n-1}$, then a linear equation in the variables can be written as

$$\sum_{i=0}^{n-1} a_i x_i = c,$$

for some real constants a_i, $i = 0, 1, \ldots, n-1$ and c. Such an equation specifies a line in n-dimensional Euclidean space.

Note: Linear vs. Affine

Technically, what we call *linear equations* do not satisfy the requirement for a function to be *linear* unless the constant term is 0. When the constant term is nonzero, such equations are said to be *affine*. However, we will use the usual convention of referring to these as *linear equations*.

Now suppose that we have two different equations for lines in 2-D space, as shown in the following equations:

$$y = 3 - x$$
$$y = 4x - 2.$$

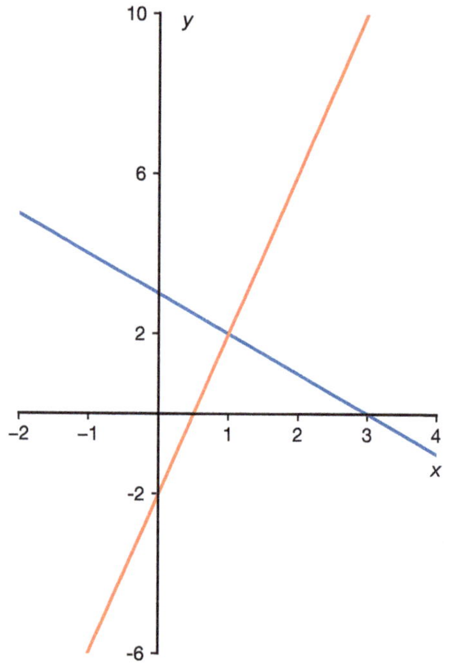

Fig. 4.2: Two linear equations.

The lines representing these equations are shown in Fig. 4.2. As can be seen from the figure, the lines intersect at the point $(1,2)$. Another way to interpret this is that the point $(1,2)$ is the only point that satisfies both these equations. When interpreted together like this, we call the two equations a *system of linear equations*:

DEFINITION

system of linear equations

A collection of linear equations on a common n-dimensional space (typically $\mathbb{R}^n$) that are interpreted together, typically with the purpose of finding the subset of the space that satisfies all of the equations simultaneously.

A system of linear equations can be written concisely using matrices and vectors. To do this, we have to allow vectors (and, more generally, matrices) to contain variables. Let's see how this works by writing each of the linear equations above as the dot product of a coefficient vector and a variable vector. Start by rewriting the equations with the variables on the left-hand side:

$$x + y = 3$$
$$-4x + y = -2$$

Create a variable vector $[x, y]^T$. Then the first equation can be written as

$$\begin{bmatrix} 1 & 1 \end{bmatrix} \begin{bmatrix} x \\ y \end{bmatrix} = 3,$$

and the second equation has the same form,

$$\begin{bmatrix} -4 & 1 \end{bmatrix} \begin{bmatrix} x \\ y \end{bmatrix} = -2.$$

We have two dot products involving the same right-hand vector, and we can express these concisely using matrix-vector multiplication as

$$\begin{bmatrix} 1 & 1 \\ -4 & 1 \end{bmatrix} \begin{bmatrix} x \\ y \end{bmatrix} = \begin{bmatrix} 3 \\ -2 \end{bmatrix}.$$

Now consider a general system of m equations in n variables $x_0, x_1, \ldots, x_{n-1}$:

$$a_{0,0}x_0 + a_{0,1}x_1 + \ldots + a_{0,n-1}x_{n-1} = b_0$$
$$a_{1,0}x_0 + a_{1,1}x_1 + \ldots + a_{0,n-1}x_{n-1} = b_1$$
$$\vdots$$
$$a_{m-1,0}x_0 + a_{m-1,1}x_1 + \ldots + a_{m-1,n-1}x_{n-1} = b_{m-1}.$$

The ith equation can be written as a dot product of the form

$$\begin{bmatrix} a_{i,0} & a_{i,1} & \cdots & a_{i,n-1} \end{bmatrix} \begin{bmatrix} x_0 \\ x_1 \\ \vdots \\ x_{n-1} \end{bmatrix} = b_i.$$

Let **A** be the coefficient matrix

$$\begin{bmatrix} a_{0,0} & a_{0,1} & \cdots & a_{0,n-1} \\ a_{1,0} & a_{1,1} & \cdots & a_{0,n-1} \\ \vdots & & & \\ a_{m-1,0} & a_{m-1,1} & \cdots & a_{m-1,n-1} \end{bmatrix},$$

and let

$$\mathbf{x} = \begin{bmatrix} x_0 \\ x_1 \\ \vdots \\ x_{n-1} \end{bmatrix}$$

be a vector of variables. Let

$$\mathbf{b} = \begin{bmatrix} b_0 \\ b_1 \\ \vdots \\ b_{m-1} \end{bmatrix}.$$

Then we can write this system of linear equations concisely as

$$\mathbf{Ax} = \mathbf{b}.$$

We will call the vector **b** the *result vector*, but it is also sometimes referred to as simply the *right-hand side*.

4.1.1   Types of Solution Sets

Given a system of linear equations represented by $\mathbf{Ax} = \mathbf{b}$, a numerically valued vector $\mathbf{v}$ is a *solution* to the system of linear equations if the system holds for $\mathbf{x} = \mathbf{v}$; i.e.,

$$\mathbf{Av} = \mathbf{b}.$$

Such a solution is not necessarily unique, nor does every system of linear equations necessarily have any solutions. In general, the solution set of a system of linear equations may consist of:

- one unique solution,

- many solutions, or

- no solutions.

Let's build some intuition about these different cases, starting with two-dimensional examples. We have already seen an example of a system of equations with one solution, so in the following examples, we consider systems with many or no solutions:

Example 4.1: System of Linear Equations with Many Solutions

Consider the equations

$$\begin{aligned} 2x - y &= -3 \\ -6x + 3y &= 9. \end{aligned}$$

To plot these, let's rewrite each equation in the form $y = mx + b$, which we do by first moving the x variable terms to the right-hand side:

$$\begin{aligned} -y &= -2x - 3 \\ 3y &= 6x + 9. \end{aligned}$$

Next, divide the first equation by -1 and the second equation by 3 to get

$$\begin{aligned} y &= 2x + 3 \\ y &= 2x + 3. \end{aligned}$$

Both equations define the same line, as shown in Fig. 4.3. The solution to each equation is the set of points shown along the line in the figure. Thus, any of the points on the line $y = 2x + 3$ is a valid solution to the system of equations. There are an infinite number of such solutions.

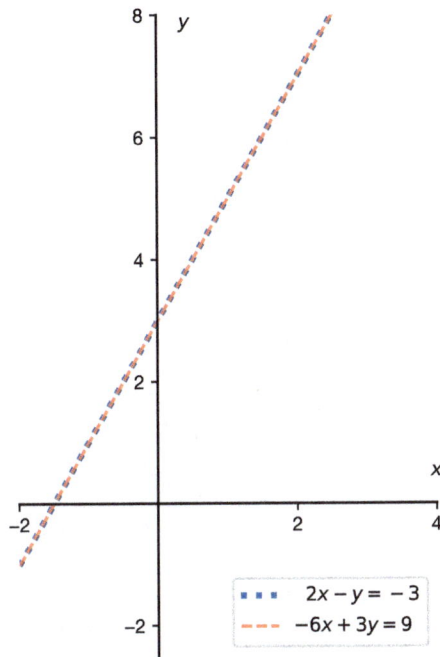

Fig. 4.3: Example of two linear equations that have many solutions.

Example 4.2: System of Linear Equations with No Solutions

Consider the equations

$$2x - y = -3$$
$$2x - y = 0.$$

Writing these in slope-intercept form, we have

$$y = 2x - 3$$
$$y = 2x + 0.$$

These equations have the same slopes but different y-intercepts. A graph of the lines that represent the solutions to these equations is shown in Fig. 4.4.

Because these lines have the same slopes, they are parallel and thus never intersect. Thus, there are no points that can simultaneously solve these equations. We say that this system of equations is *inconsistent*:

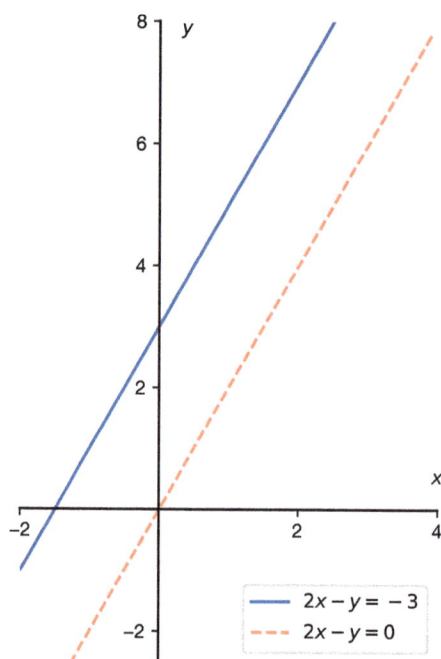

Fig. 4.4: Plot of two linear equations with no solutions.

DEFINITION

inconsistent (system of equations)

A system of equations is *inconsistent* if there are no points that simultaneously satisfy all of the equations.

Example 4.3: Another System of Linear Equations with No Solutions

Consider the set of equations shown below:

$$x + y = 3$$
$$-4x + y = -2$$
$$6x + y = 3.$$

A graph of these lines is shown in Fig. 4.5.

This set of equations has no solution because the three sets of lines do not intersect at a common point. If we take any two lines, then they intersect at a single point. The first two equations are the same as in our original two-equation example, and thus intersect at $(1, 2)$. From the graph, we see that the first and third equations intersect at $(0, 3)$, and the second and third intersect $(0.5, 0)$.

Since there is no common intersection, there is no solution to this set of equations. This system is *overdetermined*:

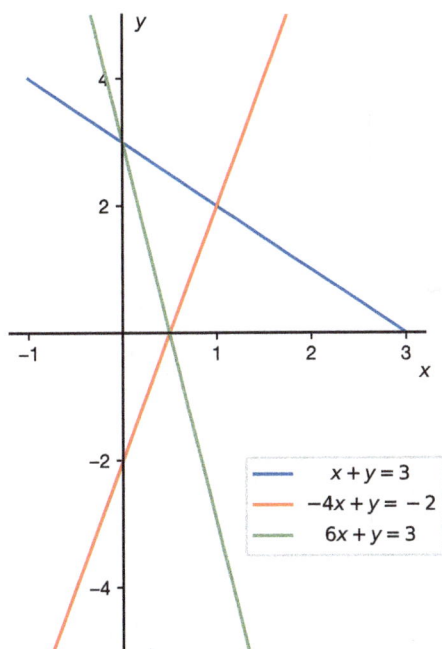

Fig. 4.5: Graph of system of three linear equations with no solution.

DEFINITION

overdetermined (system of equations)

A system of equations is *overdetermined* if there are more equations than unknowns.

An overdetermined system of equations will often be inconsistent and have no solutions (unless some equations can be written as linear combinations of the other equations – we explore this more in Section 4.2).

Example 4.4: Three-Dimensional System with Many Solutions

Consider the set of equations

$$x_1 + x_2 = 1$$
$$x_2 + x_3 = 2.$$

This system of equations is *underdetermined*:

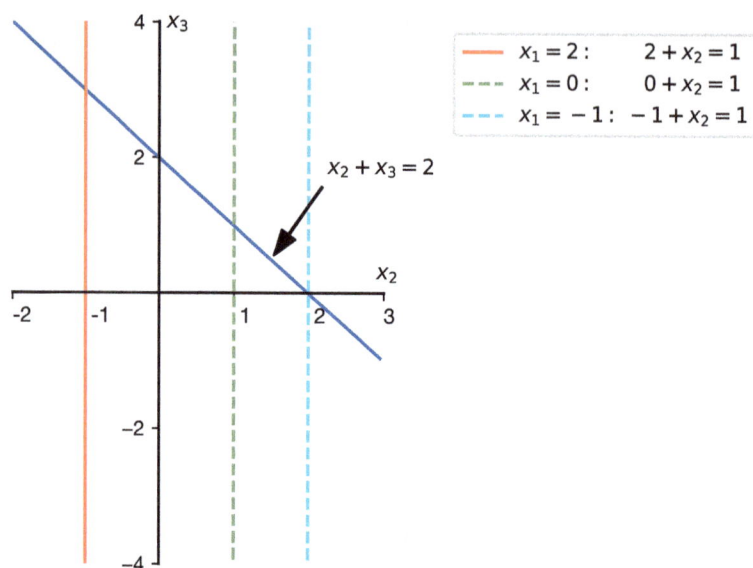

Fig. 4.6: Example solution sets for system of underdetermined equations in three variables.

DEFINITION

underdetermined (system of equations)

A system of equations is *underdetermined* if there are fewer equations than unknowns.

A system of underdetermined equations always has an infinite number of solutions.

To see that this is true for this set of equations, suppose we choose any value of x_1. Then we are left with two equations in two unknowns, and we can find a solution to the set of equations. Since we can do this for every value of x_1, there are an infinite number of solutions. Fig. 4.6 shows the solutions to the linear equations for three different values of x_1: when $x_1 = -1$, when $x_1 = 0$, and when $x_1 = 2$. When x_1 is fixed to a particular value, then the first equation can be put in the form $x_2 = c$ for some constant value c that depends on the value to which x_1 has been fixed. If we plot these two equations with x_2 on the horizontal axis and x_3 on the vertical axis, then the first equation is a vertical line for each value of x_1.

Fig. 4.6 shows the vertical lines that correspond to the first equation for each of the three different values of x_1. It also includes the line for the second equation. From the figure, we see that this system of equations has a unique solution **for each different value of x_1**. This system of equations has a different solution for each value of x_1 and therefore has an infinite number of solutions. Some of the solutions, corresponding to the three values of x_1 listed above, are $(2, -1, 3)$, $(0, 1, 1)$, and $(-1, 2, 0)$.

4.1.2 Solving Systems of Linear Equations through Row Operations

In the remainder of this section, we will consider systems of n linear equations in n variables with a single unique solution. We will solve such a system and learn how we can apply matrix techniques to facilitate solving the system. Consider the system of three equations in three variables shown below:

$$4x_0 - 3x_2 = -1$$
$$-2x_0 + 3x_1 + x_2 = -4$$
$$3x_1 - 4x_2 = -15.$$

This is an example of a *critically determined* system of equations:

> **DEFINITION**
>
> **critically determined (system of equations)**
>
> A system of equations is *critically determined* if the number of equations equals the number of unknowns.

Such a system can be solved algebraically by using linear combinations of the equations to eliminate variables. Let's see one example of how this can be done and how this connects with operations to the matrix representation.

Consider first eliminating the variable x_0 from the second equation. We can do this by multiplying the first row by $1/2$ and adding it to the second equation:

$$+\frac{1}{2}\begin{bmatrix} -2x_0 & + & 3x_1 & + & x_2 = -1 \\ 4x_0 & & & - & 3x_2 = -4 \end{bmatrix}$$

gives
$$3x_1 - \frac{1}{2}x_2 = -\frac{9}{2}.$$

Consider the effect of this operation on the system of equations expressed in matrix form, $\mathbf{Ax} = \mathbf{b}$. The original form of the system of equation is

$$\begin{bmatrix} 4 & 0 & -3 \\ -2 & 3 & 1 \\ 0 & 3 & -4 \end{bmatrix} \begin{bmatrix} x_0 \\ x_1 \\ x_2 \end{bmatrix} = \begin{bmatrix} -1 \\ -4 \\ -15 \end{bmatrix}.$$

Let's define the corresponding coefficients matrix $\mathbf{A}$ and results vector $\mathbf{b}$ as variables using NumPy:

```
# Be sure to include a decimal point after one of the numbers so that we are
# not restricted to integer values when we start manipulating this array
A = np.array([[ 4.,  0, -3],
              [ -2,  3,  1],
              [  0,  3, -4]])

b= np.array([[-1.],
             [-4 ],
             [-15]])
```

The operation of adding $1/2$ of the first equation to the second equation can be implemented by

- adding $1/2$ of row 0 of **A** to row 1 of **A**, and

- adding $1/2$ of row 0 of **b** to row 1 of **b**.

Let A2 and b2 be the modified matrix and vector:

```
A2 = A.copy()
A2[1] = A2[1] + 0.5* A2[0]

print(A2)
```

```
[[ 4.   0.  -3. ]
 [ 0.   3.  -0.5]
 [ 0.   3.  -4. ]]
```

```
b2 = b.copy()
b2[1] = b2[1] + (1/2)* b2[0]
print(b2)
```

```
[[ -1. ]
 [ -4.5]
 [-15. ]]
```

Comparing the matrix equation with A2 and b2 to the algebraic equation, we see that they are the same. It is tedious to have to manipulate the rows of **A** and **b** separately. To simplify our work, we can concatenate the columns of these matrices into an *augmented matrix*:

DEFINITION

augmented matrix

For a system of linear equations that can be expressed in matrix form $\mathbf{Ax} = \mathbf{b}$, the *augmented matrix* $(\mathbf{A}|\mathbf{b})$ is the matrix created by concatenating the columns of **A** and **b**. It is used to simplify the simultaneous manipulation of **A** and **b**, which is usually done to simplify solving this system of equations.

Let the augmented matrix be denoted by $(\mathbf{A}|\mathbf{b})$, which, from the original form of our example system of equations, is

$$\left[\begin{array}{ccc|c} 4 & 0 & -3 & -1 \\ -2 & 3 & 1 & -4 \\ 0 & 3 & -4 & -15 \end{array}\right].$$

We use a vertical bar to indicate the boundary between the portion of the matrix corresponding to **A** and that corresponding to **b**.

NumPy provides the `np.hstack()` command to horizontally stack matrices and vectors with the same number of rows. Let's refer to the augmented matrix in Python as `Ab`. The starting form of `Ab` is thus

```
Ab= np.hstack( (A,b) )
print(Ab)
```

```
[[  4.   0.  -3.  -1.]
 [ -2.   3.   1.  -4.]
 [  0.   3.  -4. -15.]]
```

Then we can perform the equivalent manipulation as above as follows:

```
Ab[1] = Ab[1] + 1/2*Ab[0]
print(Ab)
```

```
[[  4.   0.  -3.   -1. ]
 [  0.   3.  -0.5  -4.5]
 [  0.   3.  -4.  -15. ]]
```

Adding a linear combination of rows to another is one example of a *row operation*.

Important!
! Linear operations on the rows change the form of the system of linear equations but do not change the solution set.

After this first row, we can eliminate the variable x_1 from the third equation by subtracting the second equation from the third equation. Equivalently, we can subtract row 1 of the augmented matrix from row 2:

```
Ab[2] = Ab[2] - Ab[1]
print(Ab)
```

```
[[  4.   0.  -3.   -1. ]
 [  0.   3.  -0.5  -4.5]
 [  0.   0.  -3.5 -10.5]]
```

The equivalent set of equations is

$$4x_0 - 3x_2 = -1$$
$$3x_1 + (-0.5)x_2 = -4.5$$
$$-3.5x_2 = 10.5 .$$

After these manipulations, the augmented matrix shown above is said to be in *row echelon form*:

DEFINITION

row echelon form (REF)

A matrix is in row echelon form if

1. The first nonzero entry in each row, called the *leading coefficient*, is to the **right** of the leading coefficient in any row above it, and

2. Any all-zero rows are at the bottom of the matrix; any rows with nonzero coefficients appear above the all-zero rows.

Note:

Some books require the leading coefficients to be 1 for a matrix to be in row echelon form.

LU Decomposition

The REF form of $(\mathbf{A}|\mathbf{b})$ is an upper triangular matrix; let's call it $\mathbf{U}$ to distinguish it from the matrix $(\mathbf{A}|\mathbf{b})$ that we started with. Since we created this upper triangular matrix using linear combinations of the rows, there must be a matrix $\mathbf{L}$ such that $\mathbf{LU}$ is equal to the original matrix $(\mathbf{A}|\mathbf{b})$. It can be shown that the matrix $\mathbf{L}$ is a lower-triangular matrix.

The form $\mathbf{LU}$ is called the *LU decomposition* or *LU factorization* of $(\mathbf{A}|\mathbf{b})$. If row swaps are required to get a matrix in REF form, then the matrix can be factored as $\mathbf{PLU}$, where $\mathbf{P}$ is a permutation matrix, which performs row swaps. The (P)LU-factorization of a matrix can be found using SciPy's `scipy.linalg.lu()` function. Links to additional resources for the LU-decomposition are available at la4ds.net/4-1.

We can solve the system of equations for our example by iteratively solving the equations from bottom to top, where at each step, we substitute the values of the variables found in the previous step.

We start by solving the last equation for x_2, which we can do by dividing the equation by -3.5:

$$-3.5x_2 = -10.5$$
$$x_2 = \frac{-10.5}{-3.5}$$
$$= 3.$$

We can perform the equivalent operation on the augmented matrix by dividing the last row by -3.5:

```
Ab[2] = Ab[2] / -3.5
Ab
```

```
array([[ 4. ,  0. , -3. , -1. ],
       [ 0. ,  3. , -0.5, -4.5],
       [-0. , -0. ,  1. ,  3. ]])
```

Multiplying (or dividing) a row by a constant is a second type of row operation.

We can use the solution to equation 2 (the last equation) to solve the next-to-last equation in two different ways. In the first approach, we substitute the -3 for x_2 in equation 1 to reduce that equation to one unknown:

$$3x_1 + -0.5x_2 = -4.5$$
$$3x_1 + -0.5(3) = -4.5$$
$$3x_1 + = -3 \quad \text{(by adding 1.5 to each side)}$$
$$x_1 = -1 \quad \text{(dividing both sides by 3)}$$

That approach is not simple to apply to our augmented matrix.

The second approach is to add a weighted version of equation 2 to equation 1 to eliminate the variable x_2 in equation 1. If we multiply equation 2 by 0.5 and then add it to equation 1, the x_2 coefficient will be eliminated. In algebra, this looks like

$$3x_1 + -0.5x_2 = -4.5$$
$$+0.5\left(x_2 = 3 \right)$$
$$\text{gives} \qquad 3x_1 = -3 \, .$$

If we add $1/2$ of row 2 to row 1 of `Ab`, we get the equivalent result:

```
Ab[1] = Ab[1] + 0.5*Ab[2]
Ab
```

```
array([[ 4.,  0., -3., -1.],
       [ 0.,  3.,  0., -3.],
       [-0., -0.,  1.,  3.]])
```

Then by dividing row 1 by 3, we can get that $x_1 = -1$:

```
Ab[1] = Ab[1] / 3
Ab
```

```
array([[ 4.,   0.,  -3.,  -1.],
       [ 0.,   1.,   0.,  -1.],
       [-0.,  -0.,   1.,   3.]])
```

We can see that the same strategy will work for row 0. We can eliminate the x_2 variable by adding row 2 multiplied by 3:

```
Ab[0] = Ab[0] + 3*Ab[2]
Ab
```

```
array([[ 4.,   0.,   0.,   8.],
       [ 0.,   1.,   0.,  -1.],
       [-0.,  -0.,   1.,   3.]])
```

Dividing row 0 by 4 yields the solution:

```
Ab[0] = Ab[0]/4
Ab
```

```
array([[ 1.,   0.,   0.,   2.],
       [ 0.,   1.,   0.,  -1.],
       [-0.,  -0.,   1.,   3.]])
```

So $x_0 = 2$.

Note that the portion of `Ab` that corresponds to the coefficient matrix is an identity matrix. This matrix is said to be in *reduced row echelon form*:

DEFINITION

reduced row echelon form (RREF)

A matrix is in **reduced** row echelon form if it is in row echelon form and:

1. The leading coefficients are all 1.

2. For any column containing a leading coefficient, that leading coefficient is the only nonzero value in that column.

Note:

The fact that the matrix is already in row echelon form implies that all of the coefficients *below* a leading coefficient (in the same column) are zero. Condition 2 implies that all of the coefficients *above* a leading coefficient are also zero.

Any matrix may be put into row echelon form through row operations. In addition to the row operations described above, we can also swap the rows representing any pair of equations (called a *row swap*). Thus, we can define row operations as follows:

DEFINITION

row operations (matrix)

The following operations on the rows of a matrix are called *row operations*:

1. Adding any linear combination of the other rows to a row

2. Multiplying or dividing a row by a constant

3. Swapping two rows

For an augmented matrix that represents a system of linear equations, row operations do not change the solution set. In addition, the RREF is unique. As in our example, we can perform row operations to find the solution of a system of linear equations (when a unique solution exists).

Gaussian elimination, or *Gauss-Jordan elimination* is an algorithm to systematically put a matrix in row-reduced echelon form:

DEFINITION

Gaussian elimination,
Gauss-Jordan elimination

A systematic algorithm for transforming a matrix into reduced row echelon form. The algorithm iterates down the rows, performing row operations to get the matrix into row echelon form with leading coefficients equal to 1. The algorithm then iterates back up the rows to eliminate any variables above a leading coefficient through adding an appropriate linear combination of the lower row with that leading coefficient.

The basic method of Gaussian elimination follows the example above. Readers who wish to learn how to perform Gaussian elimination by hand can find many tutorials online. Instead of performing Gaussian elimination by hand, we will use Python. NumPy does not have a method to find the RREF of a NumPy array. Instead, we can use another useful library called SymPy, which has a `Matrix` class with an `rref()` method. We start by importing the `Matrix` class and making our augmented matrix into a SymPy Matrix object:

```
from sympy import Matrix

Ab2 = np.hstack( (A,b) )
M = Matrix(Ab2)
```

To get the reduced row-echelon form of the matrix M, we can call `M.rref()`. By default `M.rref()` returns the RREF of M and a tuple of pivot columns. For our purpose, we will only use the RREF, so we can pass the keyword argument `pivots=False` to suppress that output:

```
M.rref(pivots=False)
```

$$\begin{bmatrix} 1 & 0 & 0 & 2.0 \\ 0 & 1 & 0 & -1.0 \\ 0 & 0 & 1 & 3.0 \end{bmatrix}$$

Example 4.5: Solving Two Equations in Two Unknowns Using RREF

Let's return to the original set of equations that we started with in this section:

$$y = 3 - x$$
$$y = 4x - 2 \ .$$

These equations are shown in Fig. 4.2.

Let's formulate these as a matrix equation and solve. The matrix equation form is:

$$\begin{bmatrix} 1 & 1 \\ -4 & 1 \end{bmatrix} \begin{bmatrix} x \\ y \end{bmatrix} = \begin{bmatrix} 3 \\ -2 \end{bmatrix} .$$

Then the augmented matrix is

```
Ab2 = np.array([
  [ 1, 1, 3],
  [-4, 1, -2]
])
```

We can use SymPy to find the RREF for Ab2:

```
M2 = Matrix(Ab2)
M2.rref(pivots=False)
```

$$\begin{bmatrix} 1 & 0 & 1 \\ 0 & 1 & 2 \end{bmatrix}$$

The solution is $x = 1$, $y = 2$, which matches the intersection of the two lines in Fig. 4.2.

4.1.3 Calculating the Determinant of a Matrix using Row Echelon Form

Recall from Section 3.5 that there is no simple formula for the determinant of a general $n \times n$ matrix. However, one of the useful properties of the determinant is that the determinant of a triangular matrix is the product of the diagonal elements. Let $\mathbf{R_A}$ be the REF form of $\mathbf{A}$

found using only linear combinations of the rows. Then it can be shown that $\det \mathbf{A} = \det \mathbf{R_A}$. Sometimes we can simplify the mathematics of finding the REF by swapping rows of the matrix. If row swaps are also used in finding the REF and there are a total of k row swaps, then $\det \mathbf{A} = (-1)^k \det \mathbf{R_A}$. However, if the rows are also scaled, then the determinant will also be affected by these scaling factors. Such scaling is commonly used to make the first non-zero entry be 1 or to make them integer (such as can occur with SymPy's `echelon_form()` method). I am not going to consider this case because it is largely irrelevant. The only reason to use the REF to find the determinant is if a computational tool is not available, and in this case, you can calculate the determinant before applying any row scaling.

Example 4.6: Determinant of a 3×3 Matrix

Consider the matrix $\mathbf{A}$ from above. Based on the REF we found for $(\mathbf{A}|\mathbf{b})$, an REF form of $\mathbf{A}$ is

$$\begin{bmatrix} 4 & 0 & -3 \\ 0 & 3 & -0.5 \\ 0 & 0 & -3.5 \end{bmatrix}.$$

Thus $\det \mathbf{A} = (4)(3)(-3.5) = -42$. Let's check by using NumPy to compute the determinant of the original $\mathbf{A}$ matrix:

```
import numpy.linalg as la
la.det(A)
```

```
-42.00000000000001
```

4.1.4 Summary

For the two examples of solving a system of equations through matrix operations, the number of equations was equal to the number of variables (i.e., the system was not overdetermined or underdetermined), and there was a single solution. In the next section, we consider other cases that can occur with systems of linear equations.

Terminology review and self-assessment questions

Interactive flashcards to review the terminology introduced in this section and self-assessment questions are available at la4ds.net/4-1, which can also be accessed using this QR code:

4.2 Working with Systems of Linear Equations Using Matrices and Vectors – Part 2

In Section 4.1, we introduced systems of linear equations and showed examples of solving critically determined systems (with n equations in n variables) for which the system has a unique solution. In this section, we consider some other cases that we may encounter with systems of linear equations.

4.2.1   Overdetermined Systems of Linear Equations

Recall that an overdetermined system of equations has more equations than unknowns. Consider the following system of equations:

$$4x_0 - 3x_2 = -1$$
$$-2x_0 + 3x_1 + x_2 = -4$$
$$3x_1 - 4x_2 = -15$$
$$x_1 + 3x_2 - 3x_2 = -10$$

The first three equations are the same as the three-equation system of equations that we solved via the RREF in the previous section. Now there is an additional equation, for a total of 4 equations in 3 variables. Thus, this is an overdetermined set of equations.

Let's see what the implications of this are, starting with creating the augmented matrix for the matrix equation:

```python
import numpy as np

Ab = np.array([[ 4.,    0,   -3,   -1],
               [ -2,    3,    1,   -4],
               [  0,    3,   -4,  -15],
               [  1,    3,   -3,  -10]])
```

The RREF for this matrix is:

```python
from sympy import Matrix
M = Matrix(Ab)
M.rref(pivots=False)
```

$$\begin{bmatrix} 1 & 0 & 0 & 2.0 \\ 0 & 1 & 0 & -1.0 \\ 0 & 0 & 1 & 3.0 \\ 0 & 0 & 0 & 0 \end{bmatrix}$$

The first three rows of the RREF are exactly the same as when we considered the system consisting of the first three equations. The final row of the RREF is all zeros and corresponds to

$$0 \cdot x_0 + 0 \cdot x_1 + 0 \cdot x_2 = 0$$
$$0 = 0.$$

This equation is always true and provides no information about the variables. From the first three rows, we get the same solution as when we considered the system of three equations: $x_0 = 2$, $x_1 = -1$, and $x_2 = 3$. So, the fourth equation seemingly had no impact. Let's see why this occurred.

Consider if we add the first three equations together. We get

$$2x_0 + 6x_1 - 6x_2 = -20.$$

Compare that with the fourth equation: the fourth equation is equal to $1/2$ times the sum of the first three equations. We say that the fourth equation is a *linear combination* of the other three equations.

It will be easier to express the concept of a linear combination in terms of the equivalent rows of the augmented matrix:

DEFINITION

linear combination

For a matrix M, row m_{k*} is a *linear combination* of the other rows if there are constants c_i such that

$$\mathbf{m}_{k*} = \sum_{i \neq k} c_i \mathbf{m}_{i*}.$$

An equivalent definition applies to columns.

Let $(\mathbf{A}|\mathbf{b})_{i*}$ denote the ith row of the augmented matrix. Then for this augmented matrix, we have

$$(\mathbf{A}|\mathbf{b})_{3*} = \frac{1}{2}(\mathbf{A}|\mathbf{b})_{0*} + \frac{1}{2}(\mathbf{A}|\mathbf{b})_{1*} + \frac{1}{2}(\mathbf{A}|\mathbf{b})_{2*}.$$

The fact that we are expressing $(\mathbf{A}|\mathbf{b})_{3*}$ in terms of the other rows is a choice that arose from the order of equations. It is not unique. We could rewrite the linear combination to express any of the rows in terms of the other three rows. For instance, if we solve for $\frac{1}{2}(\mathbf{A}|\mathbf{b})_{0*}$, we get

$$\frac{1}{2}(\mathbf{A}|\mathbf{b})_{0*} = (\mathbf{A}|\mathbf{b})_{3*} - \frac{1}{2}(\mathbf{A}|\mathbf{b})_{1*} - \frac{1}{2}(\mathbf{A}|\mathbf{b})_{2*}$$

$$(\mathbf{A}|\mathbf{b})_{0*} = 2(\mathbf{A}|\mathbf{b})_{3*} - (\mathbf{A}|\mathbf{b})_{1*} - (\mathbf{A}|\mathbf{b})_{2*}.$$

Let's check this expression using Python:

```python
print(f'{"Ab[0] =":>24} {Ab[0]}')
print(f'{"2Ab[3] - Ab[1] -Ab[2] =":>24} {2*Ab[3] - Ab[1] -Ab[2]}')
```

```
                 Ab[0] = [ 4.  0. -3. -1.]
 2Ab[3] - Ab[1] -Ab[2] = [ 4.  0. -3. -1.]
```

When some row of a matrix can be written as a linear combination of the other rows of the matrix, the rows of the matrix are linearly dependent:

DEFINITION

linearly dependent (vectors)

A set of vectors $\{\mathbf{a}_0, \mathbf{a}_1, \ldots, \mathbf{a}_{k-1}\}$ are *linearly dependent* if any of these vectors can be written as a linear combination of the other. An equivalent condition is that there exist nonzero constants $\beta_0, \beta_1, \ldots, \beta_{k-1}$ such that

$$\beta_0 \mathbf{a}_0 + \beta_1 \mathbf{a}_1 + \ldots + \beta_{k-1} \mathbf{a}_{k-1} = \mathbf{0}.$$

Using the second condition, we can show that for any of the linearly dependent vectors, *any* of the vectors can be written as a linear combination of the other vectors.

A set of vectors that are not linearly dependent are called *linearly independent*:

DEFINITION

linearly independent (vectors)

A set of vectors that is not linearly dependent. For a set of vectors $\{\mathbf{a}_0, \mathbf{a}_1, \ldots, \mathbf{a}_{k-1}\}$ and scalar variables $\beta_0, \beta_1, \ldots, \beta_{k-1}$, the only solution to the equation

$$\beta_0 \mathbf{a}_0 + \beta_1 \mathbf{a}_1 + \ldots + t a_{k-1} \mathbf{a}_{k-1} = \mathbf{0}$$

is if $\beta_0 = \beta_1 = \ldots = \beta_{k-1} = 0$.

The maximum number of linearly independent columns of a matrix is always equal to the maximum number of linearly independent rows of a matrix. We call this number the *rank* of the matrix:

DEFINITION

rank (of a matrix)

Given a matrix $\mathbf{M}$, the *rank* is equal to the maximum number of linearly independent columns (or, equivalently, the maximum number of linearly independent rows). It is denoted by $\text{rank}\,\mathbf{M}$.

The maximum possible rank for a $m \times n$ matrix is $\min(m, n)$. If a matrix has the maximum possible rank, it is said to be *full rank*:

DEFINITION

full rank (matrix)

A matrix with the maximum possible rank, which is equal to the minimum of the number of rows and the number of columns.

In general, it can be difficult to tell if a set of vectors are linearly independent by inspection. A special case is a set of two nonzero vectors: they are linearly independent unless they are scaled versions of each other. Fortunately, NumPy has functions that can help us determine whether a set of vectors is linearly independent. The first step is to stack the vectors into either the rows or the columns of a matrix. Vectors are usually treated as column vectors and thus stacked in the columns of a matrix.

We can use two different functions or methods to get the rank of a matrix, depending on whether we are using SymPy or only NumPy. If we are using SymPy, then the `Matrix` object has a rank method. Thus, the rank of the SymPy matrix `M` is

```
M.rank()
```

3

NumPy and PyTorch have equivalent functions called `np.linalg.matrix_rank()` and `torch.linalg.matrix_rank()` that return the rank of a matrix. I use NumPy in the examples below. For convenience of typing, we will import `np.linalg` as `la`. Then we can get the rank of the augmented matrix `Ab` for this example as follows:

```
import numpy.linalg as la

la.matrix_rank(Ab)
```

3

Since the rank is three, but there are four rows and four columns, this matrix is not full rank. We say it is *rank deficient*:

> **DEFINITION**
>
> **rank deficient (matrix)**
> A matrix that is not full rank: its rank is smaller than the maximum possible.

For a square matrix, the determinant is related to a matrix's rank:

> **Important!**
>
> !   A square matrix **A** is full-rank if and only if it is nonsingular, meaning det **A** $\neq$ 0.

Let's check for `Ab`:

```
la.det(Ab)
```

```
0.0
```

The matrix `Ab` has zero determinant, so it is a singular matrix and does not have full rank.

Example 4.7: Overdetermined System with No Solutions

Linear equations can be overdetermined without any of the equations being linearly combinations of the others. For example, consider this system of equations from the previous section:

$$
\begin{aligned}
x + y &= 3 \\
-4x + y &= -2 \\
6x + y &= 3.
\end{aligned}
$$

The graph of these equations is shown in Fig. 4.5. The graph shows that this set of equations is inconsistent and thus has no solution. The augmented matrix for this set of equations is

```
Ab2 = np.array([
  [1, 1, 3],
  [-4, 1, -2],
  [6, 1, 3]
])
```

Then the RREF is

```
M2 = Matrix(Ab2)
M2.rref(pivots=False)
```

$$
\begin{bmatrix}
1 & 0 & 0 \\
0 & 1 & 0 \\
0 & 0 & 1
\end{bmatrix}
$$

At first, this may look reasonable, but remember that this augmented matrix represents the following linear equation in matrix form

$$
\begin{bmatrix}
1 & 0 \\
0 & 1 \\
0 & 0
\end{bmatrix}
\begin{bmatrix}
x \\
y
\end{bmatrix}
=
\begin{bmatrix}
0 \\
0 \\
1
\end{bmatrix}.
$$

In particular, the equation corresponding to the last row of the coefficient matrix and last row of the results vector is

$$
\begin{aligned}
0 \cdot x + 0 \cdot y &= 1 \\
\Rightarrow 0 &= 1,
\end{aligned}
$$

which is a contradiction. Thus, there is no solution to this system of equations.
The rank of the augmented matrix is

```
M2.rank()
```

3

```
la.matrix_rank(Ab2)
```

3

This matrix is *full rank*. We can confirm this by checking that the determinant is nonzero:

```
print(f'{la.det(Ab2):.2f}')
```

-25.00

However, we only have two unknowns. When the rank of the augmented matrix exceeds the number of unknowns, the system of equations will be inconsistent.

An overdetermined set of equations could also have an infinite number of solutions if a sufficient number of those equations are linearly dependent. In that case, the rank of the matrix will be smaller than the number of unknowns.

4.2.2 Underdetermined Systems of Linear Equations

An underdetermined system of equations has fewer equations than unknowns.

Example 4.8: Underdetermined System of Equations in Three Variables

Consider the following system of linear equations from the previous section:

$$x_1 + x_2 = 1$$
$$x_2 + x_3 = 2$$

The number of equations, and hence the maximum possible value of the rank of the augmented matrix, is 2, whereas the number of unknowns is 3. When the matrix rank is smaller than the number of equations, the system will always have an infinite number of solutions.

Let's confirm by computing the matrix rank with NumPy:

```
Ab3 = np.array([ [1, 1, 0, 2],
                 [0, 1, 1, 2] ])
la.matrix_rank(Ab3)
```

2

This is a *full rank matrix*. The RREF for this matrix is

```
M3 = Matrix(Ab3)
M3.rref(pivots=False)
```

$$\begin{bmatrix} 1 & 0 & -1 & 0 \\ 0 & 1 & 1 & 2 \end{bmatrix}$$

We are left with two linear equations in three unknowns. These two equations describe a line in three dimensions, and thus there are an infinite number of solutions.

4.2.3   Critically Determined Systems of Equations

For a critically determined system of equations, the number of equations is equal to the number of unknowns. The rank of the augmented matrix is less than or equal to the number of equations. Even when the system is critically determined, we may still have no solutions, one solution, or an infinite number of solutions.

Example 4.9: Critically Determined System with No Solutions

In Section 4.1.1, we showed that this system of equations is inconsistent:

$$2x - y = -3$$
$$2x - y = 0.$$

A graph of the lines that represent the solutions to these equations is shown in Fig. 4.4. For these equations, the rank of the augmented matrix is

```
M4 = Matrix([ [2, -1, -3],
              [2, -1,  0] ])
M4.rank()
```

2

which is equal to the number of variables. In this case, there will not be an infinite number of solutions, but we still have to check the RREF to see if there is a solution:

```
M4.rref(pivots=False)
```

$$\begin{bmatrix} 1 & -\frac{1}{2} & 0 \\ 0 & 0 & 1 \end{bmatrix}$$

Again, the last equation corresponds to $0 = 1$, so there is no solution to this system.

For the case of critically determined systems of linear equations, it is more common to find the rank of the coefficients matrix $\mathbf{A}$ instead of the augmented matrix $(\mathbf{A}|\mathbf{b})$. There are several reasons for this.

First, for the system to have a unique solution, rank$(\mathbf{A})$ must be equal to the number of variables. If not, then we can perform row operations on the augmented matrix to get all zeros in the coefficients portion of the augmented matrix in all but rank$(\mathbf{A})$ rows. Then there are two possibilities:

1. For every row with all zeros coefficients, there is a zero in the results portion of the row. Thus, these rows correspond to the equation $0 = 0$, which is always true. Thus, these rows can be ignored; they correspond to rows in the original $\mathbf{A}$ matrix that can be expressed as linear combinations of other rows. Since we have fewer equations in the RREF than we have unknowns, the system has an infinite number of solutions.

2. For some row with all zero coefficients, there is a nonzero value in the results portion of that row. Such a row corresponds to an equation of the form $0 = c$, where $c \neq 0$, which is always false. In this case, the equations are inconsistent, and there is no solution.

In either case, if rank$(\mathbf{A})$ is smaller than the number of variables, then there is not a unique solution. For a critically determined system, the number of columns of $\mathbf{A}$ (corresponding to the number of variables) is equal to the number of rows of A (corresponding to the number of equations), and $\mathbf{A}$ is a square matrix. Thus, $\mathbf{A}$ must be full rank for the system to have a unique solution.

Let's check rank$(\mathbf{A})$ for Example 4.9:

```
A4 = M4[:2,:2]
A4.rank()
```

```
1
```

We see that the rank is 1, but we have two variables. Thus, $\mathbf{A}$ is not full rank, and we do not have a unique solution. By using the rank of $\mathbf{A}$ instead of the rank of $(\mathbf{A}|\mathbf{b})$, we were able to identify that the system did not have a unique solution without having to find the RREF.

Secondly, for a critically determined system of equations, the $\mathbf{A}$ matrix will be square, and we can easily check if $\mathbf{A}$ is full rank (i.e., nonsingular) by checking if its determinant is nonzero.

A third reason for using the rank of the coefficients matrix is that it gives an answer that does not depend on the results vector $\mathbf{b}$. In many scenarios, we are interested in finding solutions to a system of linear equations for which the coefficients matrix depends on the structure of some system and is fixed, but the results vector varies over time. If **A** is of full rank, we can find the solution for **x** regardless of the values in **b**.

Example 4.10: Critically Determined System with Infinite Solutions

Consider again the following equations first introduced in Section 4.1.1:

$$2x - \ y = -3$$
$$-6x + 3y = \ \ 9$$

In Section 4.1.1, we showed that the solution set to both equations is the same line, as shown in Fig. 4.3. In addition, we showed that the second equation is equal to the first equation multiplied by -3; equivalently the first equation is equal to the second equation divided by -3. The two equations are linearly dependent; therefore, we expect the matrix rank will only be 1. Let's check:

```
M5 = Matrix([ [ 2, -1, -3],
              [-6,  3,  9] ])
M5.rank()
```

1

The RREF for this matrix is

```
M5.rref(pivots=False)
```

$$\begin{bmatrix} 1 & -\frac{1}{2} & -\frac{3}{2} \\ 0 & 0 & 0 \end{bmatrix}$$

We see that we are left with one equation in two unknowns, and thus the solution is all of the points that satisfy this equation. There are an infinite number of such solutions, corresponding to all of the points in the line shown in the figure.

The rank of the coefficients matrix is

```
A5 = M5[:2,:2]
A5.rank()
```

1

Again, we see that A5 is not full rank, and therefore this system of equations does not have a unique solution. Instead of finding the rank, we could have checked the determinant. The SymPy Matrix class has a method for computing the determinant:

```
A5.det()
```

0

Since the determinant is zero, A5 is singular and cannot be of full rank.

Example 4.11: Critically Determined System in Two Unknowns with a Unique Solution

Consider again the following system of equations from Section 4.1:

$$x + y = 3$$
$$-4x + y = -2$$

A plot of the solution sets for these equations is show in Fig. 4.2. The rank and determinant of the coefficients matrix are shown below:

```
A6 = np.array([ [ 1, 1],
                [-4, 1] ])

print(f'rank: {la.matrix_rank(A6)}')
print(f'determinant: {la.det(A6): .1f}')
```

```
rank: 2
determinant:  5.0
```

The coefficients matrix is full rank, and this is confirmed by the nonzero determinant. We found that the solution was $(1, 2)$ in Section 4.1.

Example 4.12: Critically Determined System in Three Unknowns with One Solution

In Section 4.1.2, we solved the system

$$4x_0 - 3x_2 = -1$$
$$-2x_0 + 3x_1 + x_2 = -4$$
$$3x_1 - 4x_2 = -15$$

using row operations to find the RREF.

Let's check the rank and determinant of the **A** matrix:

```python
A7 = np.array([ [4, 0, -3],
                [-2, 3, 1],
                [0, 3, -4] ])

print(f'rank: {la.matrix_rank(A7)}')
print(f'determinant: {la.det(A7): .1f}')
```

```
rank: 3
determinant: -42.0
```

This matrix has *full rank* and is nonsingular. We found the solution to be $(2, -1, 3)$ in Section 4.1.2.

4.2.4 Summary

In this section, we considered examples of many different types of situations that can occur with overdetermined, underdetermined, and critically determined systems of linear equations. I introduced the concepts of linear combinations of equations and rows/columns of matrices, and I explained how these define the rank of a matrix. Then I showed how the rank of the coefficients matrix or augmented matrix can help us determine whether a system of equations has a unique solution.

Terminology review and self-assessment questions

Interactive flashcards to review the terminology introduced in this section and self-assessment questions are available at la4ds.net/4-2, which can also be accessed using this QR code:

4.3 Matrix Inverses and Solving Systems of Linear Equations

Consider again the problem of solving a critically-determined system of linear equations in matrix form,

$$\mathbf{Ax} = \mathbf{b}.$$

We have previously shown how to solve such a system using the reduced row echelon form (RREF). One problem with solving a system of linear equations using the RREF is that if **b** changes, then we have to find the RREF again. Or do we?

Suppose that **A** is a square matrix with full rank (i.e., is nonsingular). Then the system of equations has a unique solution. The row operations needed to transform **A** into the identity matrix do not depend on the results vector at all. Therefore, all we need to do to solve the system of equations for different **b** is to keep track of how the sequence of row operations affect **b**. Let's do this for the two-dimensional example that has a unique solution, which we previously solved using SymPy to find the RREF of the augmented matrix in Example 4.5. Here, we will find the RREF step-by-step to see what insight it gives to the general problem.

Example 4.13: Tracking Row Operations Used to Solve a Two Variable System

The following equations have a unique solution:

$$x + y = 3$$
$$-4x + y = -2$$

The augmented matrix is

```python
import numpy as np

A = np.array([
  [1, 1],
  [-4, 1]])

b= np.array([
  [3],
  [-2]])

Ab = np.hstack( (A,b) )
```

To get the RREF, the first step is to add 4 times row 0 to row 1:

```python
Ab[1]  += 4*Ab[0]
Ab
```

```
array([[ 1,  1,  3],
       [ 0,  5, 10]])
```

Note that this changed the result vector from $[3, -2]^T$ to $[3, 10]^T$, where the change in b_1 is attributable to adding 4 times row 0 to row 1. The entry b_0 is unchanged. We can implement the effect of the row operation on the **b** vector using matrix multiplication as

$$\begin{bmatrix} 1 & 0 \\ 4 & 1 \end{bmatrix} \mathbf{b}.$$

Let's verify this

```
np.array([
    [1, 0],
    [4, 1]]) @ b
```

```
array([[ 3],
       [10]])
```

The second row operation to get the augmented matrix in RREF form is to multiply row 1 by 1/5, which yields

```
Ab[1] = Ab[1] / 5
Ab
```

```
array([[1, 1, 3],
       [0, 1, 2]])
```

Again, we can implement this operation using matrix multiplication. Let's first show this matrix multiplication as occurring separately from the previous one:

$$\begin{bmatrix} 1 & 0 \\ 0 & 1/5 \end{bmatrix} \begin{bmatrix} 1 & 0 \\ 4 & 1 \end{bmatrix} \mathbf{b}.$$

Let's verify that this works:

```
np.array([
    [1, 0],
    [0, 1/5]]) @ \
np.array([
    [1, 0],
    [4, 1]]) @ \
b
```

```
array([[3.],
       [2.]])
```

Finally, we need to subtract row 1 from row 0:

```
Ab[0] -= Ab[1]
Ab
```

```
array([[1, 0, 1],
       [0, 1, 2]])
```

Implementing this as a matrix, we get the following:

$$\begin{bmatrix} 1 & -1 \\ 0 & 1 \end{bmatrix} \begin{bmatrix} 1 & 0 \\ 0 & 1/5 \end{bmatrix} \begin{bmatrix} 1 & 0 \\ 4 & 1 \end{bmatrix} \mathbf{b}.$$

Let's check:

```
np.array([
  [1, -1],
  [0, 1]]) @ \
np.array([
  [1, 0],
  [0, 1/5]]) @ \
np.array([
  [1, 0],
  [4, 1]]) @ \
b
```

```
array([[1.],
       [2.]])
```

We can carry out all the matrix multiplications, except for the multiplication with **b**, to find a single matrix that multiplies **b**. That matrix is

```
np.array([
  [1, -1],
  [0, 1]]) @ \
np.array([
  [1, 0],
  [0, 1/5]]) @ \
np.array([
  [1, 0],
  [4, 1]])
```

```
array([[ 0.2, -0.2],
       [ 0.8,  0.2]])
```

Thus, the answer for any **b** is equal to

$$\begin{bmatrix} 1/5 & -1/5 \\ 4/5 & 1/5 \end{bmatrix} \mathbf{b}.$$

There is an easy way to find the matrix that pre-multiplies **b** during the Gaussian Elimination process. Instead of applying Gaussian Elimination to $(\mathbf{A}|\mathbf{b})$, consider what happens if we perform Gaussian Elimination on $(\mathbf{A}|\mathbf{I})$.

Below, we perform the same three row operations described above, but now we perform them using the $(\mathbf{A}|\mathbf{I})$ matrix. First, add 4 times row 0 to row 1.

```
AI = np.hstack ( (A, np.eye(2)) )
AI[1] += 4* AI[0]
AI
```

```
array([[1., 1., 1., 0.],
       [0., 5., 4., 1.]])
```

The submatrix in the rightmost two columns, which was $\mathbf{I}_2$ before the row operations, is now the same matrix we found previously for multiplying **b** in the first step.

Now divide row 1 by 5:

```
AI[1] /= 5
AI
```

```
array([[1. , 1. , 1. , 0. ],
       [0. , 1. , 0.8, 0.2]])
```

The submatrix in the right two columns is now equal to the product of the two matrices that we found through the first two steps above.

Finally, subtract row 1 from row 0:

```
AI[0] -= AI[1]
AI
```

```
array([[ 1. , 0. , 0.2, -0.2],
       [ 0. , 1. , 0.8,  0.2]])
```

The submatrix in the right three columns is exactly the matrix we need to multiply **b** to solve this equation.

Let's temporarily call this matrix **C**:

```
C = AI[:, 2:]
C
```

```
array([[ 0.2, -0.2],
       [ 0.8,  0.2]])
```

Let's see what happens if we use this matrix to left-multiply **A**:

```
np.round(C @ A, 10)
```

```
array([[ 1., -0.],
       [ 0.,  1.]])
```

We get the identity matrix! We created this matrix to implement the effects of the row operations to transform **A** into an identity matrix, so when we apply it to **A**, we get the identity matrix. We call this the *inverse* of **A**:

DEFINITION

inverse (of a matrix)

Given a $n \times n$ square matrix **A**, the *inverse* matrix (if it exists) is an $n \times n$ matrix denoted $\mathbf{A}^{-1}$, such that

$$\mathbf{A}^{-1}\mathbf{A} = \mathbf{I}.$$

For a square matrix **A**, the inverse will exist if **A** has full rank. In this case, we say that **A** is invertible:

DEFINITION

invertible (matrix)

A square matrix **A** is *invertible* if its inverse $\mathbf{A}^{-1}$ exists; this corresponds to **A** having full rank, which is true if **A** is nonsingular.

Important!

!   If **A** is an $n \times n$ matrix with full rank, we can find the inverse matrix by putting $(\mathbf{A}|\mathbf{I}_n)$ into RREF. The inverse is the submatrix consisting of columns n to $2n - 1$.

Example 4.14: Inverse of a 3×3 Matrix and Use in Solving System of Three Equations in Three Variables

Let's test this with our three-dimensional system of equations with a unique solution:

$$4x_0 - 3x_2 = -1$$
$$-2x_0 + 3x_1 + x_2 = -4$$
$$3x_1 - 4x_2 = -15$$

```
A2 = np. array([
  [4, 0, -3],
  [-2, 3, 1],
  [0, 3, -4]])


AI2 = np.hstack( (A2, np.eye(3)) )
```

```
from sympy import Matrix

M2 = Matrix(AI2)
M2r = M2.rref(pivots=False)
```

Thus, the inverse matrix is

```
A2inv = M2r[:,3:]
A2inv
```

$$
\begin{bmatrix}
0.357142857142857 & 0.214285714285714 & -0.214285714285714 \\
0.19047619047619 & 0.380952380952381 & -0.0476190476190476 \\
0.142857142857143 & 0.285714285714286 & -0.285714285714286
\end{bmatrix}
$$

Let's check

```
A2inv @ A2
```

$$
\begin{bmatrix}
1.0 & 0 & 0 \\
0 & 1.0 & 0 \\
0 & 0 & 1.0
\end{bmatrix}
$$

This confirms that the matrix A2inv is the inverse of A2. We can get the solution to the system of equations by multiplying A2inv by the results vector b:

```
b2 = np.array([
  [-1],
  [-4],
  [-15]])


A2inv @ b2
```

$$\begin{bmatrix} 2.0 \\ -1.0 \\ 3.0 \end{bmatrix}$$

This matches the answer we previously found.

In general, for a critically determined system of equations with matrix $\mathbf{A}$, where $\mathbf{A}$ has full rank, we can solve the system by left-multiplying both sides by $\mathbf{A}^{-1}$:

$$\mathbf{Ax} = \mathbf{b}$$
$$\mathbf{A}^{-1}\mathbf{Ax} = \mathbf{A}^{-1}\mathbf{b}$$
$$\mathbf{Ix} = \mathbf{A}^{-1}\mathbf{b}$$
$$\mathbf{x} = \mathbf{A}^{-1}\mathbf{b}.$$

We can find the inverse of a square, full-rank matrix using NumPy's np.linalg.inv() function or PyTorch's torch.linalg.inv() function. Note that PyTorch's method requires a tensor of floats, so multiply an integer tensor by 1.0 if necessary. Here is an example using NumPy:

```
import numpy.linalg as la

print('A =\n', A)
print('A^(-1) =\n', la.inv(A))

print()
print('A2 = \n', A2)
print('A2^(-1) = \n', la.inv(A2))
```

```
A =
 [[ 1  1]
  [-4  1]]
A^(-1) =
 [[ 0.2 -0.2]
  [ 0.8  0.2]]
```

(continues on next page)

(continued from previous page)

```
A2 =
 [[ 4  0 -3]
  [-2  3  1]
  [ 0  3 -4]]
A2^(-1) =
 [[ 0.35714286  0.21428571 -0.21428571]
  [ 0.19047619  0.38095238 -0.04761905]
  [ 0.14285714  0.28571429 -0.28571429]]
```

When using a SymPy Matrix, we can use the `inv()` method to get the inverse. One advantage is that the inverse is often expressed in a nicer form (using fractions instead of reals):

```
M = Matrix(A)
M.inv()
```

$$\begin{bmatrix} \frac{1}{5} & -\frac{1}{5} \\ \frac{4}{5} & \frac{1}{5} \end{bmatrix}$$

```
M2 = Matrix(A2)
M2.inv()
```

$$\begin{bmatrix} \frac{5}{14} & \frac{3}{14} & -\frac{3}{14} \\ \frac{4}{21} & \frac{8}{21} & -\frac{1}{21} \\ \frac{1}{7} & \frac{2}{7} & -\frac{2}{7} \end{bmatrix}$$

4.3.1   More on Determinants and Inverses

As previously mentioned, for a square matrix to be full rank, it must be nonsingular. Thus, a simple criterion for a square matrix to be invertible is:

> **Important!**
>
> **! A matrix is invertible (full rank) if and only if its determinant is nonzero.**

To see one reason this is true, recall that the determinant of a matrix product is the product of the determinants. Suppose that $\mathbf{A}$ is invertible but the determinant of $\mathbf{A}$ is zero. Then

$$\det\left(\mathbf{A}^{-1}\mathbf{A}\right) = \det\left(\mathbf{A}^{-1}\right)\det\left(\mathbf{A}\right)$$
$$\det\left(\mathbf{I}\right) = \det\left(\mathbf{A}^{-1}\right)(0)$$
$$1 = 0,$$

where the last step follows from the facts that the determinant of the identity matrix is 1 and the product of zero and anything is zero. Thus, this leads to a contradiction.

The same relation implies that if $\det \mathbf{A} \neq 0$, then

$$\det \mathbf{A}^{-1} = \frac{1}{\det \mathbf{A}}.$$

When using the determinant to check if $\mathbf{A}$ is invertible, be aware that the results from NumPy may be subject to numerical errors from floating point operations in the computer. For example, consider the following coefficients matrix:

```
A3 = np.array([
    [4, -4, -1],
    [12, 4, -7],
    [4, 12, -5]])
la.det(A3)
```

```
-2.8421709430404045e-14
```

The value returned is nonzero, but that is due to computational errors. The matrix is not invertible, as we can see from its RREF:

```
M3 = Matrix (A3)
M3.rref(pivots=False)
```

$$\begin{bmatrix} 1 & 0 & -\frac{1}{2} \\ 0 & 1 & -\frac{1}{4} \\ 0 & 0 & 0 \end{bmatrix}$$

Another advantage of using SymPy's Matrix object is that determinants are calculated using fractions and do not suffer from the same types of computational errors:

```
M3.det()
```

0

4.3.2   Special Cases

I recommend you use computational tools to find matrix inverses when required.

Important!

!

You should always check the determinant of a matrix to be sure the matrix is nonsingular before trying to invert it.

For students that are not allowed to use a computer on an exam, many scientific calculators can calculate matrix inverses. However, it is helpful to know how to find the matrix inverse in some special cases:

1. **General Full-rank 2×2 Matrices**
   For a 2×2 matrix $\mathbf{A}$ of the form

$$\mathbf{A} = \begin{bmatrix} a & b \\ c & d \end{bmatrix},$$

the inverse is

$$\mathbf{A}^{-1} = \frac{1}{\det A} \begin{bmatrix} d & -b \\ -c & a \end{bmatrix}.$$

The relation can be summarized as follows:

- swap the entries on the diagonal,

- negate the entries off the diagonal, and

- divide by the determinant of the original matrix.

2. **Diagonal Matrices**
   A diagonal matrix is invertible if and only if all of the entries on the diagonal are nonzero; otherwise, the determinant is zero. For a diagonal matrix $\mathbf{A}$ of the form

$$\mathbf{A} = \begin{bmatrix} a_{00} & 0 & \cdots & 0 \\ 0 & a_{11} & \cdots & 0 \\ \vdots & \vdots & \ddots & \vdots \\ 0 & 0 & \cdots & a_{n-1,n-1} \end{bmatrix},$$

the inverse is the diagonal matrix of inverses,

$$\mathbf{A}^{-1} = \begin{bmatrix} 1/a_{00} & 0 & \cdots & 0 \\ 0 & 1/a_{11} & \cdots & 0 \\ \vdots & \vdots & \ddots & \vdots \\ 0 & 0 & \cdots & 1/a_{n-1,n-1} \end{bmatrix}.$$

3. **Orthogonal Matrices**
   If $\mathbf{U}$ is an orthogonal matrix, then its columns are orthogonal, which implies that

$$\mathbf{U}^T \mathbf{U} = \mathbf{I}.$$

Thus, for an orthogonal matrix, its inverse is equal to its transpose: $\mathbf{U}^{-1} = \mathbf{U}^T$.

4.3.3 Properties of Inverse

Three simple properties of matrix inverses are:

1. **Inverse works on either side:** Recall that matrix multiplication is not generally commutative. Even for square matrices **A** and **B** of the same size, **AB** is not generally equal to **BA**. However, if **A** is a square matrix with inverse $\mathbf{A}^{-1}$, then $\mathbf{A}^{-1}\mathbf{A} = \mathbf{A}\mathbf{A}^{-1} = \mathbf{I}$.

2. **Inverse of matrix transpose:** If **A** is a square matrix with inverse $\mathbf{A}^{-1}$, then $(\mathbf{A}^T)^{-1} = (\mathbf{A}^{-1})^T$.

3. **Inverse of a matrix product:** If **A** and **B** are invertible matrices, then $(\mathbf{AB})^{-1} = \mathbf{B}^{-1}\mathbf{A}^{-1}$.

4.3.4 Summary and Discussion

In this section, I introduced matrix inverses and showed how they can be used to solve systems of linear equations. Finding the matrix inverse using RREF requires on the order of n^3 operations for an $n \times n$ matrix, but more efficient algorithms can reduce this complexity to less than $n^{2.4}$. Using the matrix inverse is a practical method to solve small systems of equations, especially if the solution has to be found for multiple result vectors. In practical systems with large matrices, it is generally recommended to avoid using the matrix inverse for this purpose. The reasons for this are:

1. Finding the matrix inverse can be more complex than direct methods of solving the equations.

2. Using the matrix inverse is more likely to result in larger numerical errors than direct methods of solving the equations.

3. Many practical matrix equations are sparse and can be stored efficiently, but the matrix inverse is not sparse and therefore may require too much storage space. In addition, operations with that matrix will be inefficient because the data cannot be kept in the CPU cache.

Terminology review and self-assessment questions

Interactive flashcards to review the terminology introduced in this section and self-assessment questions are available at la4ds.net/4-3, which can also be accessed using this QR code:

4.4 Application to Eigenvalues and Eigenvectors

We can use the information covered in the previous sections to show how to solve by hand for the eigenvalues and eigenvectors of a matrix. I also show how to decompose a non-singular matrix into a matrix product involving the modal matrix and the matrix with the eigenvalues on its diagonal. We also investigate the relationship between the determinant and the eigenvalues.

4.4.1 Solving for Eigenvalues and Eigenvectors

Consider again equation (3.2), which defines an eigenvector-eigenvalue pair, and rewrite it slightly as shown:

$$\mathbf{Mv} = \lambda \mathbf{u} \tag{4.1}$$

$$\Rightarrow \mathbf{Mu} - \lambda \mathbf{u} = \mathbf{0}$$

$$\Rightarrow (\mathbf{M} - \lambda \mathbf{I})\,\mathbf{u} = \mathbf{0}. \tag{4.2}$$

Recall from the column interpretation of matrix multiplication in Section 3.2.4 that a matrix-vector product can be interpreted as a linear combination of the columns of the matrix, where the linear coefficients are the components of the vector. Thus, each eigenvector specifies a linear combination of the columns of $\mathbf{M} - \lambda \mathbf{I}$ that add to give the zero vector. This implies that the columns are linearly dependent. So $\mathbf{M} - \lambda \mathbf{I}$ is not full rank and is a singular matrix. Singular matrices have determinant zero, and any eigenvalue λ must satisfy

$$\det (\mathbf{M} - \lambda \mathbf{I}) = 0.$$

In practice, it is often more convenient to multiply the argument of the determinant by -1, which does not change the solution. This is called the *characteristic equation*:

> **DEFINITION**
>
> **characteristic equation,**
> **characteristic polynomial**
>
> Given a $n \times n$ matrix $\mathbf{M}$, the *characteristic equation* or *characteristic polynomial* is
>
> $$\det (\lambda \mathbf{I} - \mathbf{M}) = 0, \tag{4.3}$$
>
> which is a polynomial equation in terms of λ that can be used to solve for the eigenvalues (λ).

Although we will generally use NumPy to find the eigenvector-eigenvalue pairs of a matrix, solving for the eigenvalues of a 2×2 matrix is relatively easy, as shown in the following example:

Example 4.15: Finding Eigenvalues Using the Characteristic Equation

Consider finding the eigenvalues of the matrix $\mathbf{M}_5$. Then the argument of the determinant in the characteristic equation is

$$\begin{bmatrix} \lambda & 0 \\ 0 & \lambda \end{bmatrix} - \begin{bmatrix} 1/2 & -4 \\ -2 & 3 \end{bmatrix} = \begin{bmatrix} \lambda - 1/2 & -4 \\ -2 & \lambda - 3 \end{bmatrix}.$$

The determinant of the resulting matrix can be calculated by taking the product of the diagonal elements and subtracting the product of the off-diagonal elements:

$$(\lambda - 1/2)(\lambda - 3) - (-4)(-2) = 0$$
$$\lambda^2 - 3.5\lambda + 1.5 - 8 = 0$$
$$\lambda^2 - 3.5\lambda - 6.5 = 0.$$

We can then solve for the eigenvalues using the quadratic formula:

```
a = 1
b = -3.5
c = -6.5

print(f'Eigenvalue 1: {(-b - np.sqrt(b**2 -4*a*c))/(2*a) : .2f}')
```

```
Eigenvalue 1: -1.34
```

This value matches the first eigenvalue from `la.eig()`. Note that I chose to use the smaller of the solutions as the one called "Eigenvalue 1" here because it matches up with the order of the solutions from `la.eig()`. However, in general, the eigenvalues are not inherently ordered, and I could have called the larger solution as "Eigenvalue 1".

Let's check the other solution to the quadratic equation:

```
print(f'Eigenvalue 2: {(-b + np.sqrt(b**2 -4*a*c))/(2*a) : .2f}')
```

```
Eigenvalue 2:  4.84
```

As expected, the result is equal to the second eigenvalue reported by `la.eig()`.

When performing the calculations by hand for 2×2 matrices, a much faster approach to finding the eigenvalues is given in Section 4.4.4.

Now let's consider the problem of finding the eigenvectors given that we have the eigenvalues. Recall that eigenvalues and eigenvectors come in pairs. Thus, we solve for the ith eigenvector, $\mathbf{u}_i$, by substituting the ith eigenvalue, λ_i into (4.2) and solving

$$(\mathbf{M} - \lambda_i \mathbf{I}) \mathbf{u}_i = 0. \tag{4.4}$$

But we know from our previous work that the matrix $\mathbf{M} - \lambda_i \mathbf{I}$ is singular. Let's check for our example:

Example 4.16: Determinant of Matrix in Characteristic Equation for 2×2 Matrix

Consider again the the matrix $\mathbf{M}_5$, for which we found the eigenvalues in Example 4.15. The eigenvalues are approximately $\lambda_0 \approx -1.34$ and $\lambda_1 \approx 4.84$. Our previous work says that the matrix $\mathbf{M}_5 - \lambda_i \mathbf{I}$ is singular and so should have determinant zero. The code below verifies this using the eigenvalues `lam0` and `lam1` found above:

```
la.det(M5 - lam0*np.eye(2)),  la.det(M5 - lam1*np.eye(2))
```

```
(-1.2070945703753258e-15, -9.641907759346243e-16)
```

The equations given by (4.4) for each i are always underdetermined. This makes sense because we know that if $\mathbf{u}$ is an eigenvalue of a matrix $\mathbf{M}$, then $c\mathbf{u}$ is also an eigenvector of $\mathbf{M}$ for any $c \neq 0$. To solve (4.4), we just need to add another constraint on $\mathbf{u}_i$. Ideally, we would use $\|\mathbf{u}\| = 1$ because we usually want to get the unit-norm eigenvectors; however, that is a nonlinear constraint because it can be written in terms of the norm-squared as

$$\sum_k (u_{i,k})^2 = 1.$$

Instead, we can use a linear constraint of our choice on the values of the eigenvector, provided the choice is linearly independent of the other linear equations to be solved. For this example, I suggest to use

$$\sum_k u_{i,k} = 1.$$

Mathematically, we can replace the last row of $\mathbf{M} - \lambda_i\mathbf{I}$ with a row of ones and replace the last entry in the zero vector on the righthand side of (4.4) with a 1. Call the solution for this constraint $\tilde{\mathbf{u}}_i$. After solving, we can find the unit-norm eigenvector as $\mathbf{u}_i = \tilde{\mathbf{u}}_i/\|\tilde{\mathbf{u}}_i\|$. Let's demonstrate this using our example:

Example 4.17: Calculation of Eigenvectors for Example 2×2 Matrix

Consider again the matrix from Example 4.15. The code below calculates $N_0 = \mathbf{M} - \lambda_0\mathbf{I}$, changes the last row to ones, and then solves for $\mathbf{u}_0$ using the matrix inverse:

```
N0 = M5 - lam0*np.eye(2)
N0[1] = np.ones(2)
u0t = la.inv(N0) @ np.array( [[0, 1]] ).T
u0t
```

```
array([[0.68465844],
       [0.31534156]])
```

Let's confirm that this is an eigenvector. If it is, then if we perform component-wise division of the projected vector by the original vector, each of the elements of the results vector should be the eigenvalue:

```
print(f'lambda 0 = {lam0:.3g}')
M5 @ v0t / v0t
```

```
lambda 0 = -1.34
```

```
array([[-1.34232922],
       [-1.34232922]])
```

We can create a unit-norm eigenvector by dividing the eigenvector we found by its own norm:

```
u0t / la.norm(u0t)
```

```
array([[0.90828954],
       [0.41834209]])
```

Let's compare with the eigenvectors found via `np.eig()`. Recall that the eigenvectors are the columns of the modal matrix, which we found in using `np.eig()` in Example 3.10. The output for this matrix is:

```
U
```

```
array([[-0.90828954,  0.67752031],
       [-0.41834209, -0.73550406]])
```

The eigenvector we found is the negative of the eigenvector in the first column of the modal matrix. This can happen, since both are unit-norm eigenvectors for λ_0.

4.4.2 Eigendecomposition

Suppose we have an $n \times n$ matrix $\mathbf{M}$ with n known eigenvectors $\mathbf{u}_i$, $i = 0, 1, \ldots, n-1$. Now consider (4.1) again. Rather than computing the left-hand side for each eigenvector, we can compute the product of $\mathbf{M}$ with all of the eigenvectors using the matrix product $\mathbf{MU}$, where $\mathbf{U}$ is the modal matrix. Similarly, we can compute the right-hand side of (4.1) for all of the eigenvectors as $\mathbf{\Lambda U}$, where $\mathbf{\Lambda}$ is the diagonal matrix of eigenvalues:

$$\mathbf{\Lambda} = \begin{bmatrix} \lambda_0 & 0 & 0 & \cdots & 0 \\ 0 & \lambda_1 & 0 & \cdots & 0 \\ \vdots & \vdots & \vdots & \ddots & \vdots \\ 0 & 0 & 0 & \cdots & \lambda_{n-1} \end{bmatrix}.$$

Then the relation becomes

$$\mathbf{MU} = \mathbf{\Lambda U}. \tag{4.5}$$

Let's check that the two sides are equal for our example matrix:

Example 4.18: Checking Eigenvector Equality Using the Modal Matrix:

Consider again the Python matrix M5, whose modal matrix U5 and eigenvalue vector lam5 we found in Example 3.10. Then the following code computes first the left-hand and then the right-hand side of (4.5) using Python:

```
M5 @ U5
```

```
array([[ 1.21922359,  3.28077641],
       [ 0.56155281, -3.56155281]])
```

```
U5 @ np.diag(lam5)
```

```
array([[ 1.21922359,  3.28077641],
       [ 0.56155281, -3.56155281]])
```

The two outputs are equal, as expected.

Now suppose that we have a matrix $\mathbf{M}$ that satisfies (4.5) and that has linearly independent eigenvectors. Then $\mathbf{U}$ has full rank, and its inverse $\mathbf{U}^{-1}$ exists. Right-multiplying both sides of the equation above by $\mathbf{U}^{-1}$ yields a way to express $\mathbf{M}$ in terms of $\mathbf{U}$ and $\mathbf{\Lambda}$:

$$\mathbf{M} = \mathbf{U}\mathbf{\Lambda}\mathbf{U}^{-1}.$$

This is called the *eigendecomposition* of $\mathbf{M}$:

DEFINITION

**eigendecomposition,
matrix diagonalization**

Suppose $\mathbf{M}$ is a real $n \times n$ matrix with modal matrix $\mathbf{U}$ and eigenvalue matrix $\mathbf{\Lambda}$. If $\mathbf{U}$ has full rank, then the *eigendecomposition* (also known as the diagonalization) of $\mathbf{M}$ is the factorization

$$\mathbf{M} = \mathbf{U}\mathbf{\Lambda}\mathbf{U}^{-1}.$$

Example 4.19: Eigendecomposition of Example 2×2 Matrix

Let's confirm that eigendecomposition works for our 2×2 example matrix, $\mathbf{M}_5$:

```
Lam5 = np.diag(lam5)

U5 @ Lam5 @ la.inv(U5)
```

```
array([[ 0.5, -4. ],
       [-2. ,  3. ]])
```

The result is the original matrix $\mathbf{M}_5$.

We will use eigendecomposition in Section 6.4 to find an alternate representation for data that allows us to extract the portions of the data that are "most important" in a certain sense.

4.4.3 Relating Eigenvalues to Matrix Determinant

Let $\mathbf{M}$ be a $n \times n$ matrix that has eigendecomposition

$$\mathbf{M} = \mathbf{U}\mathbf{\Lambda}\mathbf{U}^{-1},$$

where $\mathbf{U}$ has the same dimensions as $\mathbf{M}$ (i.e., $\mathbf{M}$ has n eigenvectors). One of the properties of the determinant is that the determinant of the product of matrices is the product of the determinants, so

$$\det \mathbf{M} = (\det \mathbf{U})(\det \mathbf{\Lambda})\left(\det \mathbf{U}^{-1}\right).$$

Another property of the determinant is that

$$\det \mathbf{U}^{-1} = \frac{1}{\det \mathbf{U}}.$$

Thus, we can write

$$\det \mathbf{M} = (\det \mathbf{U})\left(\frac{1}{\det \mathbf{U}}\right)(\det \mathbf{\Lambda})$$

$$= \det \mathbf{\Lambda}$$

$$= \prod_{i=0}^{n-1} \lambda_i,$$

where the last step follows from the fact that the determinant of a diagonal matrix is the product of its diagonal elements.

Example 4.20: Calculating Determinant of a Matrix from Its Eigenvalues

Consider again our 2×2 example matrix $\mathbf{M}_5$ whose eigenvalues we have calculated using NumPy and stored in the variable lam. We can easily calculate the determinant by hand as shown in Section 3.5. The result is $(0.5)(3) - (-4)(-2) = -6.5$. Let's check the product of eigenvalues and the result from la.det():

```
np.prod(lam), la.det(M5)
```

```
(-6.499999999999999, -6.499999999999999)
```

The results agree, although there is some small computational error.

So, if a matrix has an eigendecomposition, then its determinant is equal to the product of its eigenvalues. In particular:

- A matrix is singular if and only if its determinant is zero. Thus, a matrix is singular if and only if it has a zero eigenvalue.

- A matrix is nonsingular if and only if its determinant is nonzero. Thus, a matrix is nonsingular if and only if all of its eigenvalues are nonzero.

4.4.4   Matrix Trace and a Fast Way to Find Eigenvalues of a 2×2 Matrix

Although eigenvalues should generally be found using computer techniques, like `np.eig()`, students are occasionally asked to find them by hand for small matrices. In Section 4.4.1, we showed how to find the eigenvalues of a matrix by solving the characteristic equation (4.3). However, for an $m \times m$ matrix, this will mean solving an mth degree polynomial. For the case of a 2×2 matrix, it is relatively easy to solve the resulting quadratic, but there is an easier way, which I originally learned from Grant Sanderson (the creator of the YouTube channel 3Blue1Brown, https://www.youtube.com/c/3blue1brown). I will introduce the technique in the same way that Grant does in the video "A quick trick for computing eigenvalues | Chapter 15, Essence of linear algebra" (https://www.youtube.com/watch?v=e50Bj7jn9IQ). We start with defining a very simple operation called *matrix trace*:

DEFINITION

trace (matrix)

The *trace* of a $m \times m$ matrix M, denoted $\text{tr}(M)$, is the sum of the elements on its main diagonal:

$$\text{tr}(M) = \sum_{k=0}^{m-1} m_{k,k}.$$

An interesting property of the matrix trace is that the trace of a matrix is equal to the sum of its eigenvalues. Consider a 2×2 matrix $\mathbf{M}$, and let the eigenvalues be denoted λ_0 and λ_1. Then

$$m_{00} + m_{11} = \lambda_0 + \lambda_1.$$

It will be convenient to divide both sides by 2, which implies that the average value (mean) of the diagonal elements is equal to the average value (mean) of the eigenvalues. Let m denote this value:

$$m = \frac{m_{00} + m_{11}}{2} = \frac{\lambda_0 + \lambda_1}{2}.$$

Recall also that the determinant of a matrix is equal to the product of the eigenvalues, $\det(M) = \lambda_0 \lambda_1$, which we call p. Then by matching terms in the solution of the characteristic equation, Sanderson shows that for a 2×2 matrix, the eigenvalues satisfy

$$\lambda = m \pm \sqrt{m^2 - p}.$$

The eigenvalues are symmetric around the mean of the diagonal elements of the matrix, and the spread depends on that mean and the determinant of the matrix.

Example 4.21: Solving for Eigenvalues Using Trace and Determinant of a 2×2 Matrix

Consider again the matrix $\mathbf{M}_5$. In Example 4.20, we showed that $p = \det \mathbf{M}_5 = -6.5$. The trace is $(\mathbf{M}_5) = 0.5 + 3 = 3.5$, so $m = 1.75$. Then the eigenvalues are

$$1.75 + \sqrt{(1.75)^2 - (-6.5)} \approx 4.84, \text{ and}$$
$$1.75 - \sqrt{(1.75)^2 - (-6.5)} \approx -1.34,$$

which match the values we previously found.

Terminology review and self-assessment questions

Interactive flashcards to review the terminology introduced in this section and self-assessment questions are available at la4ds.net/4-4, which can also be accessed using this QR code:

4.5 Approximate Solutions to Inconsistent Systems of Linear Equations

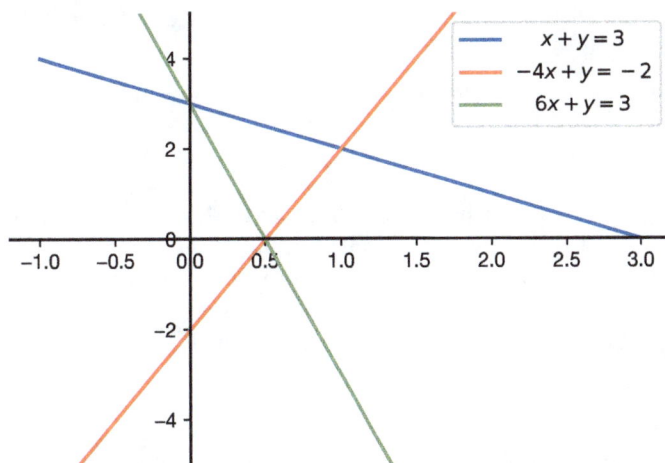

Fig. 4.7: Graph of system of three linear equations with no solution.

Consider again the three inconsistent linear equations shown in Fig. 4.7. These can be written in the matrix form

$$\begin{bmatrix} 1 & 1 \\ -4 & 1 \\ 6 & 1 \end{bmatrix} \begin{bmatrix} x \\ y \end{bmatrix} = \begin{bmatrix} 3 \\ -2 \\ 3 \end{bmatrix}. \tag{4.6}$$

More generally, consider the problem of finding the solution to a set of equations of the form

$$\mathbf{Ax} = \mathbf{b},$$

where $\mathbf{A}$ is a *tall* matrix:

DEFINITION

tall matrix

> An $m \times n$ matrix $\mathbf{A}$ is *tall* if $m > n$; i.e., the number of rows is greater than the number of columns.

Similarly, we can define a *wide* matrix:

DEFINITION

wide matrix

> An $m \times n$ matrix $\mathbf{A}$ is *wide* if $n > m$; i.e., the number of columns is greater than the number of rows.

For a matrix equation $\mathbf{Ax} = \mathbf{b}$:

- If $\mathbf{A}$ is tall, then the system of equations is overdetermined and generally has no solution. Since $\mathbf{A}$ is not square, it does not have an inverse.

- If $\mathbf{A}$ is wide, then the system of equations is underdetermined and generally has many solutions. Again, since $\mathbf{A}$ is not square, it does not have an inverse.

- If $\mathbf{A}$ is square, then the system of equations is critically determined. $\mathbf{A}$ has an inverse if the equations are linearly independent.

To understand how we can find a "good" compromise solution to a tall matrix, let's start by rewriting the matrix equation as follows:

$$\mathbf{Ax} - \mathbf{b} = \mathbf{0}.$$

If there is no solution to the left-hand side that yields all zeros, we can define the error as $\mathbf{e}(\mathbf{x}) = \mathbf{Ax} - \mathbf{b}$ and try to find a solution that minimizes the error in some sense. Since the result of $\mathbf{Ax} - \mathbf{b}$ can consist of both positive and negative errors, we cannot just add up the values. Instead, we will use our usual approach of minimizing the squared error. Since we have a vector of values, we will use the norm-squared of the error:

$$\|\mathbf{e}(\mathbf{x})\|^2 = \|\mathbf{Ax} - \mathbf{b}\|^2. \tag{4.7}$$

The solution that minimizes (4.7) is called the least-squares solution.

Recall that $\mathbf{e}(\mathbf{x})$ is a vector, and let $e_i(\mathbf{x})$ denote the ith component of the error. Then the total squared error is

$$\|\mathbf{e}(\mathbf{x})\|^2 = \sum_{i=0}^{n-1} [e_i(\mathbf{x})]^2. \tag{4.8}$$

Example 4.22: Finding the Least-Squares Solution to Three Equations in Two Variables

Consider finding the least-squares solution to (4.6). Using (4.7) and (4.8), we can write the squared error as

$$(x + y - 3)^2 + (-4x + y + 2)^2 + (6x + y - 3)^2. \qquad (4.9)$$

Thus, we want to find the values of x and y that minimize this multinomial. Let's start by visualizing the squared error as a function of x and y. To do this, in Fig. 4.8, I have plotted the solution sets to the three linear equations, and I have overlaid this with a heatmap that shows the squared error from (4.9). For each point (x, y) in the figure, the squared error from (4.9) is shown as a color, where every color maps to a numerical value. The relation between colors and values is shown in the color bar at the right of the figure. The minimum values are shown in yellow, and the yellow elliptical region is centered around approximately $(0.5, 0.8)$.

We could use `scipy.optimize.minimize()` to search for a solution or NumPy's `np.argmin()` function to search a grid for the approximate value, but we can also find the exact value using calculus, as shown below.

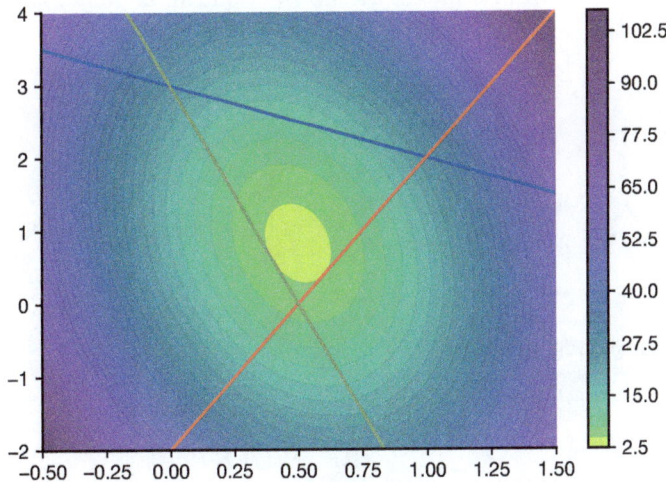

Fig. 4.8: Plot of three inconsistent linear equations overlaid with contour plot of squared error from the point to these lines.

The value of $\mathbf{x}$ that minimizes $\|\mathbf{e}(\mathbf{x})\|^2$ can be found by taking the derivatives of this equation with respect to *each* of the variables x_i and setting them equal to zero:

$$\frac{\partial}{\partial x_i} \mathbf{e}(\mathbf{x}) = 0, \quad i = 0, 1, \ldots, n - 1.$$

In the language of multi-dimensional calculus, we can express this concisely using gradient notation

$$\nabla \mathbf{e}(\mathbf{x}) = 0.$$

Without going into details of the matrix calculus, the gradient of $\mathbf{e(x)}$ is

$$\nabla \mathbf{e(x)} = 2\mathbf{A}^T \left(\mathbf{Ax} - \mathbf{b}\right).$$

Note the similarity of this form to the derivative of the single-dimensional error, $(ax - b)^2$:

$$\frac{d}{dx}(ax - b)^2 = 2a(ax - b).$$

Setting the gradient equal to zero and distributing, we have

$$\mathbf{A}^T \mathbf{Ax} - \mathbf{A}^T \mathbf{b} = 0$$
$$\Rightarrow \mathbf{A}^T \mathbf{Ax} = \mathbf{A}^T \mathbf{b}$$

These are called the *normal equations*, and the left side contains the Gram matrix, $\mathbf{A}^T \mathbf{A}$, which was introduced in Section 3.4. Note that the Gram matrix is a square matrix, and it is invertible if the columns of $\mathbf{A}$ are linearly independent. If the Gram matrix is invertible, then we can solve for the vector $\mathbf{x}$ that minimizes the squared error as[1]

$$\hat{\mathbf{x}} = \left(\mathbf{A}^T \mathbf{A}\right)^{-1} \mathbf{A}^T \mathbf{b}.$$

This the least-squares (LS) solution.

For the example shown in Fig. 4.8, the LS solution can be found as shown below.

```
A = np.array([
    [1,  1],
    [-4, 1],
    [6,  1]
])

b = np.array([3, -2, 3])

xLS = la.inv(A.T @ A) @ A.T @ b
print(xLS)
```

```
[0.5        0.83333333]
```

This result matches the value found from the contour plot.

The matrix $(A^T A)^{-1} A^T$ is called the *pseudoinverse* of $\mathbf{A}$:

---

[1]As with other equations in this book involving matrix inverses, the normal equations are usually solved via other more numerically stable methods.

DEFINITION

pseudoinverse,
Moore-Penrose pseudoinverse

For an $m \times n$ matrix real $\mathbf{A}$ with $m > n$ (i.e., $\mathbf{A}$ is tall) and linearly independent columns, the *Moore-Penrose pseudoinverse* of $\mathbf{A}$ is denoted by $\mathbf{A}^\dagger$ and given by

$$\mathbf{A}^\dagger = (\mathbf{A}^T \mathbf{A})^{-1} \mathbf{A}^T.$$

Example 4.23: Calculating the Pseudoinverse Using NumPy

The following code calculates the pseduoinverse for the coefficients matrix of (4.6) using the definition and directly using the `pinv()` function from NumPy's linear algebra module[1]:

---
[1]PyTorch has an equivalent `torch.linalg.pinv()` function.

```
print( la.inv(A.T @ A) @ A.T)
print( np.round(la.pinv(A), 10) )
```

```
[[ 0.         -0.1         0.1        ]
 [ 0.33333333  0.43333333  0.23333333]]
[[ 0.         -0.1         0.1        ]
 [ 0.33333333  0.43333333  0.23333333]]
```

This process of finding $\mathbf{x}$ is called *ordinary least squares*:

DEFINITION

ordinary least squares (OLS)

Consider an over-constrained system of linear equations of the form $\mathbf{A}\mathbf{x} = \mathbf{b}$, where $\mathbf{A}$ and $\mathbf{b}$ are known constant matrices. *Ordinary least squares (OLS)* gives the value for $\mathbf{x}$ that minimizes the norm of the error vector $\|\mathbf{A}\mathbf{x} - \mathbf{b}\|^2$. If $\mathbf{A}^\dagger$ (the Moore-Penrose inverse for $\mathbf{A}$) exists, then the OLS solution $\mathbf{x}$ is given by

$$\hat{\mathbf{x}} = \mathbf{A}^\dagger \mathbf{b}.$$

---

pseduoinverse of a wide matrix

If $\mathbf{A}$ is a wide matrix with linearly independent rows, then the pseudoinverse is defined as $\mathbf{A}^\dagger = \mathbf{A}^T (\mathbf{A}\mathbf{A}^T)^{-1}$. This form is not used in this book but is included here for reference.

Terminology review and self-assessment questions

Interactive flashcards to review the terminology introduced in this section and self-assessment questions are available at la4ds.net/4-5, which can also be accessed using this QR code:

4.6   Chapter Summary

This chapter focused on the application of matrix techniques to systems of linear equations. We saw that a system of linear equations can have no solution, one solution, or an infinite number of solutions. I demonstrated how to use matrix techniques to determine when a system of linear equations will have a unique solution and different approaches to solve for the solution(s). Finally, I showed how to find the least-squares solution to a system of over-constrained linear equations. In Chapter 5, I show the application of these techniques to problems in data fitting, including polynomial fitting and linear regression.

Access a list of key take-aways for this chapter, along with interactive flash-cards and quizzes at la4ds.net/4-6, which can also be accessed using this QR code:

5

Exact and Approximate Data Fitting

A common task in data science is finding a polynomial that represents the relation between variables in a dataset. Such representations provide information about how different factors affect each other and can be used to make predictions about scenarios that are different than those that have already been observed. In this chapter, I show how to use the techniques from Chapter 4 to find a polynomial fit to a dataset. Both exact and approximate data-fitting techniques are considered, and an application to linear regression is shown.

5.1 Exact Data Fitting with Polynomials

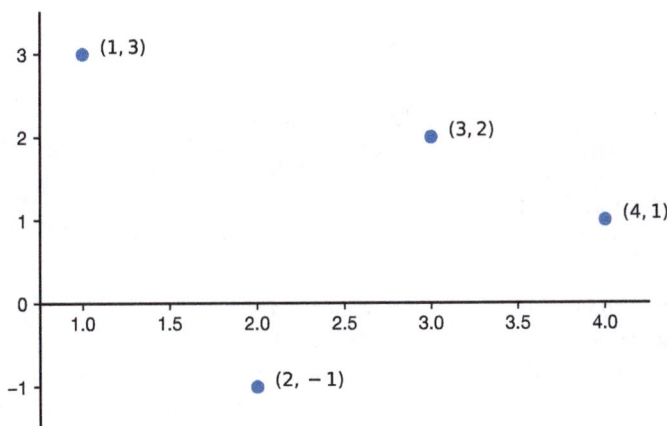

Fig. 5.1: Four data points in the real plane that are to be fit with a polynomial.

Consider the problem of finding a polynomial that passes through a set of n data points, where each data point is a point $(x_i, y_i) \in \mathbb{R}^2$ (the real plane). For example, consider finding a polynomial to fit the four data points shown in Fig. 5.1. If the data points are in the solution set for the polynomial, then we say that the polynomial *fits* the data. For example, Fig. 5.2 shows several different polynomials that fit this data. The process of finding such a polynomial from a set data points is called *polynomial fitting*. When the polynomial fits the data exactly, it is called *polynomial interpolation*.

We will show that if we have a unique set of x_i, then we can always find a polynomial of degree less than or equal to $n-1$ that can fit the n data points exactly. Let the polynomial be

$$p(x) = c_0 + c_1 x + c_2 x^2 + \ldots c_{n-1} x^{n-1}.$$

DOI: 10.1201/9781032664088-5

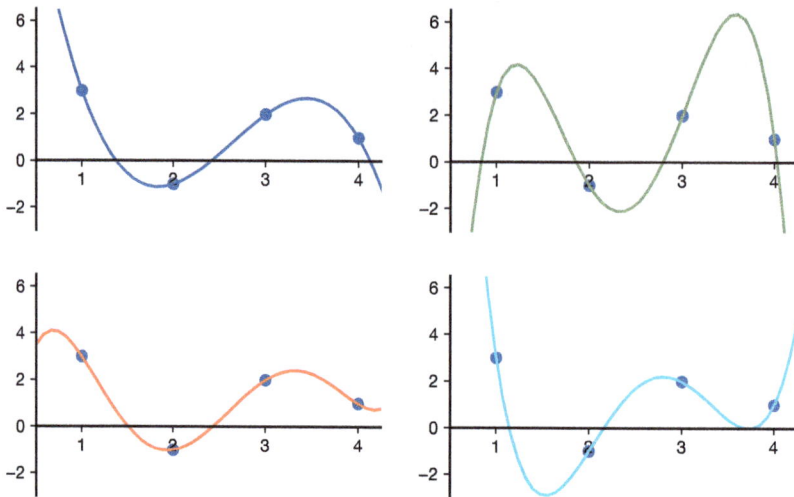

Fig. 5.2: Four data points with different polynomials that fit the data.

To find a polynomial that fits the data, we want to solve for the coefficients $c_0, c_1, \ldots, c_{n-1}$ such that the output of the polynomial for each input x_i is the corresponding value y_i:

$$p(x_0) = y_0$$
$$p(x_1) = y_1$$
$$\ldots$$
$$p(x_{n-1}) = y_{n-1}.$$

The ith equation looks like

$$y_i = c_0 + c_1 y_i + c_2 y_i^2 + \ldots c_{n-1} y_i^{n-1}. \tag{5.1}$$

At first glance, this may seem to be a set of **nonlinear** equations. But that is not the case, because all of the values for x_i and y_i are known. Thus, the powers of x_i are all **deterministic constants**. For each i, (5.1) is a linear equation in the variables $c_0, c_1, \ldots c_{n-1}$. Since we have n data points, this results in a system of n linear equations that we can put in the form

$$\mathbf{Ac} = \mathbf{y}.$$

Here, the coefficient matrix $\mathbf{A}$ is given by

$$\mathbf{A} = \begin{bmatrix} 1 & x_0 & x_0^2 & x_0^3 & \cdots & x_0^{n-1} \\ 1 & x_1 & x_1^2 & x_1^3 & \cdots & x_0^{n-1} \\ \vdots & \vdots & \vdots & \vdots & \ddots & \cdots \\ 1 & x_{n-1} & x_{n-1}^2 & x_{n-1}^3 & \cdots & x_0^{n-1} \end{bmatrix}.$$

The vector of variables $\mathbf{c}$ consists of the values of c_i that we are trying to find,

$$\mathbf{c} = \begin{bmatrix} c_0 \\ c_1 \\ \vdots \\ c_{n-1}. \end{bmatrix}$$

The results vector is

$$\mathbf{y} = \begin{bmatrix} y_0 \\ y_1 \\ \vdots \\ y_{n-1} \end{bmatrix}.$$

Provided that the set of x_i points is unique, the matrix $\mathbf{A}$ will have full rank and thus be invertible. Therefore, we can find the coefficients of the polynomial to fit the data as

$$\mathbf{c} = \mathbf{A}^{-1}\mathbf{y}.$$

Let's demonstrate this using the example data from Fig. 5.1:

Example 5.1: Exact Polynomial Fit for Four Data Points

The data values in Fig. 5.1 are $(1, 3)$, $(2, -1)$, $(3, 2)$, and $(4, 1)$. Start by creating $\mathbf{x}$ and $\mathbf{y}$ vectors from the data. When we create the $\mathbf{x}$ vector, we want to make sure it is a two-dimensional NumPy array, so that we can use `np.hstack()` to stack the powers of $\mathbf{x}$ into the coefficients matrix $\mathbf{A}$:

```python
import numpy as np

x=np.array([[ 1, 2, 3, 4 ]]).T
y=np.array([[ 3, -1, 2, 1 ]]).T
```

Since we have 4 data points, we need four powers of $\mathbf{x}$, from 0 to 3, in our $\mathbf{A}$ matrix:

```python
A=np.hstack((x**0, x**1, x**2, x**3))
A
```

```
array([[ 1,  1,  1,  1],
       [ 1,  2,  4,  8],
       [ 1,  3,  9, 27],
       [ 1,  4, 16, 64]])
```

We can check whether $\mathbf{A}$ is invertible using the determinant:

```python
import numpy.linalg as la

print(f'{la.det(A):.1f}')
```

```
12.0
```

Then the coefficients of the polynomial to fit this data are given by $\mathbf{A}^{-1}\mathbf{y}$, which yields

```
c = la.inv(A) @ y
print(c.T)
```

```
[[ 25.          -34.66666667  14.5          -1.83333333]]
```

The polynomial of degree 3 that fits this data is (approximately) $25 - 34.67x + 14.5x^2 - 1.83x^3$. Let's plot this polynomial with our data:

```python
import matplotlib.pyplot as plt

fig = plt.figure()
ax = fig.add_subplot(111)

# Plot the data
ax.scatter(x, y);


# Draw the polynomial
xl =np.linspace(-0.2, 4.2, 100)
y3 = 25 - 34.667*xl + 14.5*xl**2 - 1.8333*xl**3


plt.plot(xl, y3, 'C1')

ax.set_xlim(0.5, 4.25);
ax.set_ylim(-3, 6.5);
ax.spines['bottom'].set_position('zero');
```

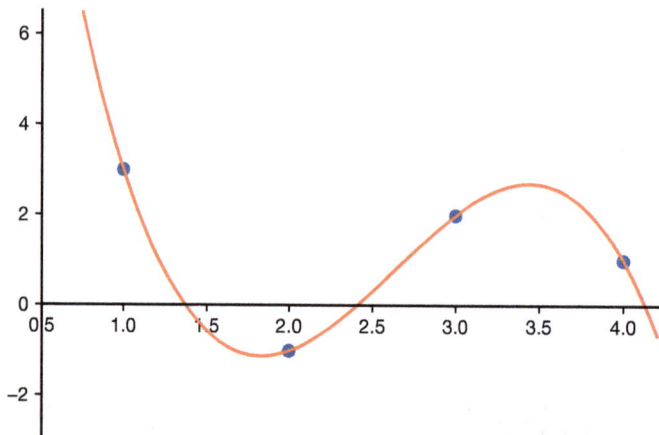

Example 5.2: Fitting Four Data Points with a Quadratic Polynomial

What happens if we try to fit the data from Example 5.1 with a quadratic polynomial instead of a cubic? The matrix of powers of the x-coordinates (the **A** matrix) in this case will consist of the first three columns of the **A** matrix from Example 5.1:

```
A2=A[:, :3]
A2
```

```
array([[ 1,  1,  1],
       [ 1,  2,  4],
       [ 1,  3,  9],
       [ 1,  4, 16]])
```

From the dimensions of A2, we know that the rank of this matrix is at most 3. We can use NumPy to verify that its rank is equal to 3:

```
la.matrix_rank(A2)
```

```
3
```

We now have four equations in three unknowns. Let's create the augmented matrix and see if these equations are consistent:

```
from sympy import Matrix

Ab2 = np.hstack( (A2, y) )
M2 = Matrix(Ab2)
M2.rref(pivots=False)
```

$$\begin{bmatrix} 1 & 0 & 0 & 0 \\ 0 & 1 & 0 & 0 \\ 0 & 0 & 1 & 0 \\ 0 & 0 & 0 & 1 \end{bmatrix}$$

The last equation corresponds to $0 = 1$, so the equations are not consistent, and there is no solution. We need a third-order polynomial to fit this data exactly.

The problem we will encounter with real data is that even if the data follows a low-order polynomial trend, noise in the data will require using a polynomial with the same order as the data to find an exact fit. Thus, in Section 5.2, we consider how to find a good polynomial fit to data when the polynomial order does not allow an exact fit.

Before we leave the topic of exact fits, let's also see what happens if the data can be fitted with a lower-order polynomial.

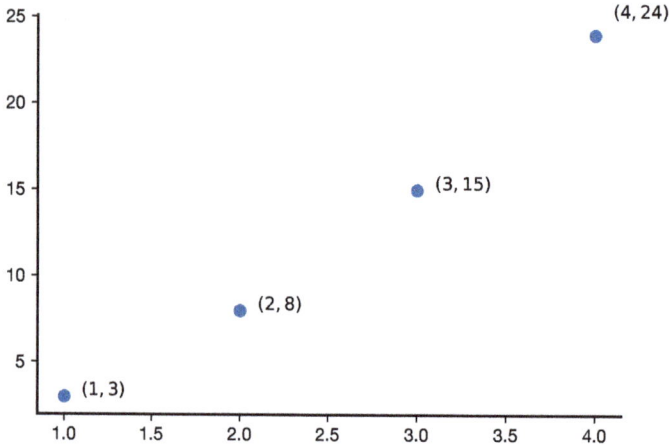

Fig. 5.3: Second example dataset consisting of four data points in the real plane.

Example 5.3: Polynomial Fit When Data Can be Fit with Lower Order Polynomial

Suppose our data consists of the following points: $(1, 3)$, $(2, 8)$, $(3, 15)$, and $(4, 24)$, as shown in Fig. 5.3. These data points are not linear, but it is hard to tell what order polynomial is required to fit them. Let's see what happens if we fit this data with a third-order polynomial. First, note that our **A** matrix is unchanged – the only part of the equation that changes is the results vector. Let's call the new results vector $\mathbf{y}_2$. It is

```
y2 = np.array([[3, 8, 15, 24]]).T
```

Then the third-order interpolating polynomial is

```
np.round(la.inv(A) @ y2, 10).T
```

```
array([[0., 2., 1., 0.]])
```

This is the polynomial $2x + x^2$. So, a quadratic is sufficient to fit this data. Let's plot it and check:

```
plt.scatter(x, y2)
plt.plot(xl, 2*xl + xl**2, 'C1');
```

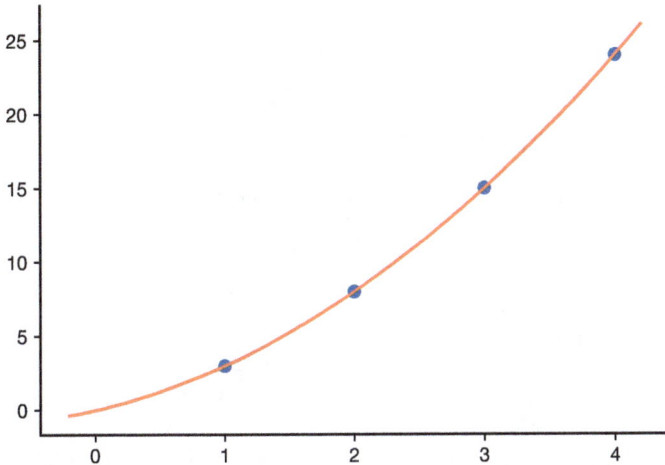

5.1.1 Summary and Discussion

In this section, I showed how to use linear algebra to find an exact polynomial fit to a sequence of points in the real plane. Given a set of n data points with different x_i, I showed how to find an interpolating polynomial of degree less than or equal to $n-1$ that fits the data exactly. Since the coefficients matrix $\mathbf{A}$ created in this process is full rank, the interpolating polynomial is *unique*. Then you may be wondering — the beginning of this section shows many different polynomials interpolating the same four data points. If the interpolating polynomial is unique, how can this be? The answer is that I took some creative liberty. To create these different example polynomials, I first added additional points outside the plotted area. I then fitted the larger datasets with higher-order polynomials. The goal of showing these different plots was to motivate the discussion of polynomial fitting.

In the next section, I show how to find a polynomial to approximately fit a set of data points when the order of the polynomial is smaller than the number of data points.

Terminology review and self-assessment questions

Interactive flashcards to review the terminology introduced in this section and self-assessment questions are available at la4ds.net/5-1, which can also be accessed using this QR code:

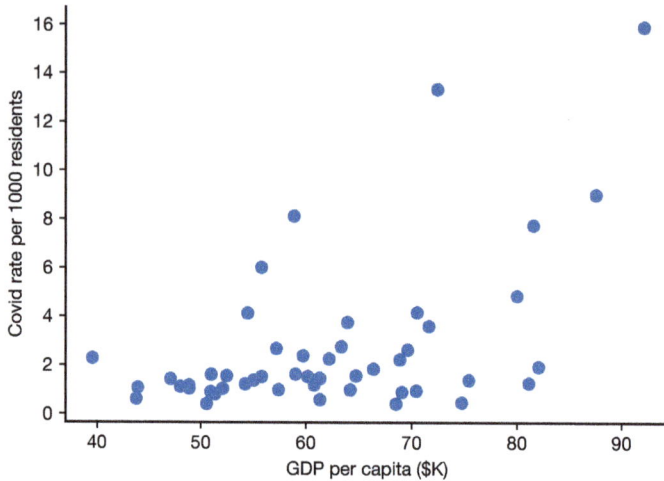

Fig. 5.4: Covid rates for GPD per capita for the 50 states of the USA.

5.2   Approximate Data Fitting

In Section 5.1, we showed that we can fit a dataset with n points with a polynomial of degree $n-1$. However, we often want to fit the data with a lower-order polynomial, which can be more useful for describing the relation among the data and for making predictions. To illustrate this, let's compare COVID-19 rates to gross domestic product (GDP) per capita for the 50 US states. I have created a CSV file at https://www.fdsp.net/data/covid-merged.csv that for each state contains:

- *state*: the state's name,

- *cases*: the number of COVID-19 cases through April 30, 2020,

- *population*: the US Census Bureau population estimate published in December 2019,

- *gdp*: the GDP as of the fourth quarter of 2019 as reported by the US Department of Commerce, and

- *urban*: the Urban Index, which is the percentage of the state's population that lives in urban areas, as determined by the US Census Bureau, as of 2010.

Fig. 5.4 shows a plot of normalized COVID rates versus GDP for the 50 states. The x-axis is the GDP per capita, reported in 1000s of dollars ($K), and the y-axis is the COVID rate per 1000 residents.

It is numerically challenging to find the correct polynomial coefficients for the exact fit using the approach presented in Section 5.1 because of computational errors. To see why such a fit would not be particularly useful even if we could find it, consider the first 10 data points and the corresponding 9th-degree interpolation polynomial, which are shown in Fig. 5.5. This polynomial is not useful because it doesn't provide a meaningful description of the relation among the data and can't be useful to make reasonable predictions – it produces negative and highly nonmonotonic predictions of the normalized COVID-19 rates. Instead, we want to figure out how to find a good **approximate** polynomial fit with a lower-order polynomial.

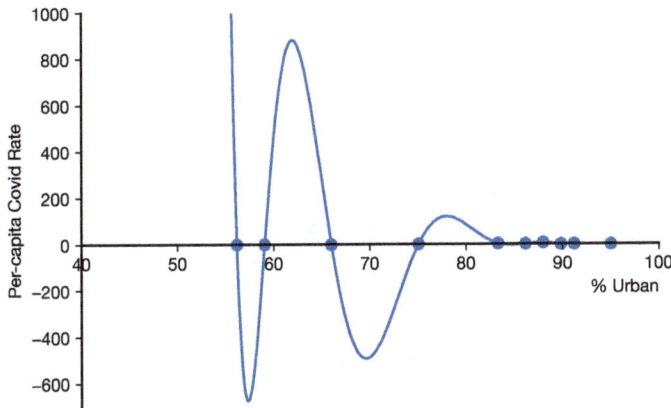

Fig. 5.5: First 10 points of COVID rate and GDP data with ninth order polynomial fit.

Before considering the general polynomial fitting problem, we start with a common problem encountered in statistics in which the data is fit with a linear equation.

5.2.1 Application to Linear Regression

Consider the relation between GDP per capita and COVID rates shown in Fig. 5.4. We often want to understand whether there is a relationship between two variables or features in a dataset, and one of the simplest ways to determine if data are related is to determine the best line of fit and see if the slope is nonzero. This is called *linear regression* and if the line of fit is chosen to minimize the total squared error to the data, this is called *ordinary least squares (OLS)*. We can use the equations for solving for the linear least-squares solution from Section 4.5 to find the OLS solution to the linear regression problem.

In linear regression, we usually classify the variables as either *explanatory variables* or *response variables*. We generally consider only one response variable at a time and classify the regression based on the number of explanatory variables:

DEFINITION

simple linear regression

Linear regression with one explanatory variable and one response variable.

DEFINITION

multiple linear regression

Linear regression with multiple explanatory variables and a single response variable.

Consider first the simple linear regression problem. We wish to find an equation $y = mx + b$ that best matches the observed data (y_i, x_i), where m is the slope of the line and b is the y-intercept. Let $\mathbf{x}$ and $\mathbf{y}$ be vectors of the observed explanatory and response values; note that the order of the data in the vectors matters to the extent that x_i and y_i must be a single observed pair of values. Let's let $\mathbf{c}_1 = [m \; b]^T$ be a vector of coefficients, which we wish to solve for. Then the resulting linear equations can be written in matrix form as

$$\mathbf{y} = \begin{bmatrix} \mathbf{x} & \mathbf{1} \end{bmatrix} \mathbf{c}_1,$$

where $\mathbf{1}$ is a ones-vector of the same length as $\mathbf{x}$ and $\mathbf{y}$.

This is usually an overdetermined set of equations. Let $\mathbf{A}_1 = [\mathbf{x} \; \mathbf{1}]$. Then we will find the least-squares solution to minimize $\|\mathbf{A}_1\mathbf{c}_1 - \mathbf{y}\|$. Comparing to Equation 4.7, we see that the OLS solution is $\mathbf{A}_1^\dagger\mathbf{y}$. Let's demonstrate this with an example.

Example 5.4: Simple Linear Regression Between GDP and COVID Rates

Consider again the data on state COVID rates and GDP per capita shown in Fig. 5.4. Let's find the OLS line of fit for this data. Let's start by importing the raw data and computing the COVID rates per 1000 people and GDP per capita in thousands of dollars:

```
import pandas as pd
covid = pd.read_csv( 'https://www.fdsp.net/data/covid-merged.csv' )

covid['gdp_norm'] = covid['gdp'] / covid['population'] * 1000;
covid['cases_norm'] = covid['cases'] / covid['population'] * 1000
```

To apply the least-square techniques described above, let's set up the variables $\mathbf{x}$, $\mathbf{y}$, and $\mathbf{A}_1$ described above for this dataset. We are interested in whether GDP (a socioeconomic factor) affects COVID rates, so we set $\mathbf{x}$ to the GDP values. The COVID rates are the response data, so we assign them to $\mathbf{y}$. The variable $\mathbf{A}_1$ is created as a matrix that contains the explanatory data $\mathbf{x}$ and a vector of 1s. Because the COVID data is treated as a row vector, it is most convenient to vertically stack the two vectors with `np.vstack()` and then transpose the resulting matrix:

```
x = covid['gdp_norm']
y =  covid['cases_norm']
A1 = np.vstack( (x ,
                np.ones_like(x)
              ) ).T
A1[:5]
```

```
array([[47.06126732, 1.        ],
       [74.73866953, 1.        ],
       [52.07219899, 1.        ],
       [43.93804236, 1.        ],
       [81.11414283, 1.        ]])
```

Then the coefficient vector can be computed as follows:

```
c1 = la.pinv(A1) @ y
print(c1)
```

```
[ 0.13790867 -5.8341525 ]
```

Let's plot the data with the OLS line of fit:

```
m = c1[0]
b = c1[1]
y_fit1 = m*x + b

plt.scatter(x, y)
plt.xlabel("GDP per capita ($K)")
plt.ylabel("Covid rate per 1000 residents");

plt.plot(x, y_fit1, color='C1');
```

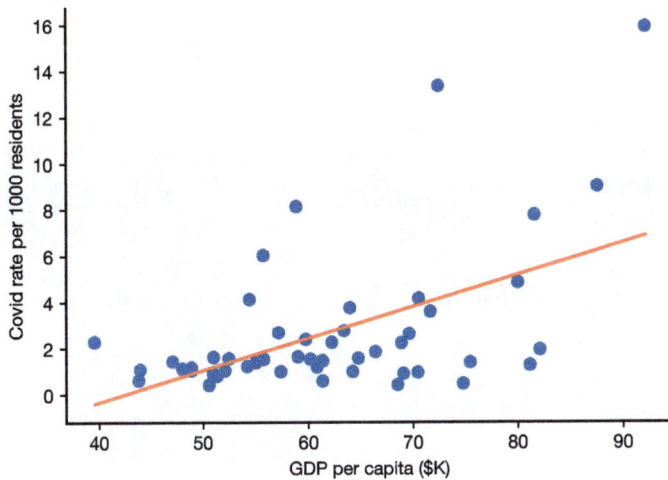

The positive slope indicates that there is a positive association between GDP and COVID rates.

The squared error is

```
la.norm(y - y_fit1)**2
```

```
354.30246867168444
```

Let's see if we can reduce that error rate by using multiple linear regression, where we will use more than one explanatory variable. Let $\mathbf{x}_0, \mathbf{x}_1, \ldots, \mathbf{x}_{k-1}$ be vectors, where vector $\mathbf{x}_i$ is the data for explanatory variable i. Then $\mathbf{A}_k$ will be

$$\mathbf{A}_k = \begin{bmatrix} \mathbf{x}_0 & \mathbf{x}_1 & \ldots & \mathbf{1} \end{bmatrix}.$$

Then the OLS solution to $\mathbf{y} = \mathbf{A}_k \mathbf{c}$ is $\mathbf{c} = \mathbf{A}_k^\dagger \mathbf{y}$. Let's test this with an example:

Example 5.5: Multiple Linear Regression with COVID Data

Consider again the COVID dataset, but now let's consider two explanatory variables. Let $\mathbf{x}_0$ be the normalized GDP data, and let $\mathbf{x}_1$ be the urban index data. Then we can create the matrix $\mathbf{A}_2$ as show below:

```
A2 = np.vstack( (covid['gdp_norm'],
                 covid['urban'],
                 np.ones_like(covid['urban'])
               ) ).T
A2[:5]
```

```
array([[47.06126732, 59.04      , 1.        ],
       [74.73866953, 66.02      , 1.        ],
       [52.07219899, 89.81      , 1.        ],
       [43.93804236, 56.16      , 1.        ],
       [81.11414283, 94.95      , 1.        ]])
```

Then the solution of the OLS problem is:

```
c2 = la.pinv(A2) @ y
c2
```

```
array([ 0.10879662,  0.04307567, -7.19829654])
```

The values in c2 are the coefficients of the best linear fit and should be interpreted as:

Normalized COVID Rate $\approx$
$$0.109(\text{Normalized GDP}) + 0.0431(\text{Urban Index}) - 7.20.$$

Let's calculate the predicted values and find the norm-squared of the error vector:

```
ecovid = Acovid @ xcovid - bcovid
la.norm(ecovid)**2
```

```
340.84175401395896
```

By using two explanatory variables, the squared error between the data and the predicted values is reduced from approximately 354.3 to 340.8.

As usual, there is an easy way to implement simple or multiple linear regression using one of the common Python data science libraries, `scikit-learn`. We will import the `LinearRegression()` function from the `sklearn.linear` submodule. `LinearRegression` returns an object with methods to perform tasks like fitting the model to the data (in this case, performing OLS) and calculating predicted values using the fitted model. When using `LinearRegression()`, it will already include a constant term, so we only need to pass the first two columns of `Ak` to the `fit()` method:

```
from sklearn.linear_model import LinearRegression

lr = LinearRegression()
lr.fit(Ak[:, :2], y);
```

The coefficients and constant term of the linear fit are given by

```
lr.coef_, lr.intercept_
```

```
(array([0.10879662, 0.04307567]), -7.198296541829917)
```

These match the ones we found using NumPy.

5.2.2 Finding an Approximate Polynomial Fit to Data

Now consider the problem of trying to find a polynomial $p(x)$ to fit a sequence of data points (x_0, y_0), (x_1, y_1), $\ldots$, (x_{n-1}, y_{n-1}). In Section 5.1, we showed how to find an exact fit with a polynomial of degree $n-1$. However, if we wish to fit the data with a polynomial of degree $m < n-1$, we may need to find an approximate fit. As before, we will use squared error as our measure of "good" and try to minimize the total squared error between the approximation and the data. Thus, if our polynomial fit is $p(x)$, then we want to choose the coefficients of $p(x)$ to minimize

$$\sum_i [y_i - p(x_i)]^2. \tag{5.2}$$

Let's reformulate this to better reveal the nature of the problem. Let n be the number of data points, and let $m < n$ be the degree of the polynomial that we want to use to approximate the data. Then the polynomial is of the form $p(x) = c_0 + c_1 x_1 + c_2 x^2 + \ldots c_{m-1} x^{m-1}$. We can calculate all of the polynomial values using the matrix equation

$$\begin{bmatrix} 1 & x_0 & x_0^2 & \ldots x_0^{m-1} \\ 1 & x_1 & x_1^2 & \ldots x_1^{m-1} \\ \ldots & \ldots & \ldots & \ddots & \ldots \\ 1 & x_{n-1} & x_{n-1}^2 & \ldots x_{n-1}^{m-1} \end{bmatrix} \begin{bmatrix} c_0 \\ c_1 \\ c_2 \\ \ldots \\ c_{m-1} \end{bmatrix}.$$

Let $\mathbf{x} = [x_0, x_1, \ldots, x_{n-1}]^T$. Then we can write the left matrix in this equation concisely as $\mathbf{A}_m = \begin{bmatrix} 1 & \mathbf{x} & \mathbf{x}^2 & \ldots & \mathbf{x}^{m-1} \end{bmatrix}$, where we use $\mathbf{x}^i$ to denote

$$\mathbf{x}^i = \begin{bmatrix} x_0^i \\ x_1^i \\ \vdots \\ x_{n-1}^i \end{bmatrix}.$$

Letting $\mathbf{y} = [y_0, y_1, \ldots, y_{n-1}]^T$, we can rewrite (5.2) as

$$\|\mathbf{A}_m \mathbf{c} - \mathbf{y}\|^2.$$

Note that $\mathbf{A}_m \mathbf{c} - \mathbf{y}$ is a linear equation, even though $p(x)$ is a polynomial; this is because it is linear in the **coefficients** of the polynomial. Thus, this is exactly the same type of ordinary least-squares problem that we encountered in Section 4.5. Provided the Gram matrix $\mathbf{A}_m^T \mathbf{A}_m$ is invertible, the solution is $\mathbf{c} = \mathbf{A}_m^\dagger \mathbf{y}$. (If the Gram matrix is not invertible, the rows of $\mathbf{A}_m$ are linearly dependent, and the size of the $\mathbf{A}_m$ matrix can be reduced by dropping a row and hence reducing the order of the polynomial by 1. This can be repeated until the Gram matrix is invertible.)

Let's apply this to find a low-order polynomial fit to the COVID data shown at the beginning of this section:

Example 5.6: Polynomial Fits to COVID and GDP Data

Let's start by approximating our data using a quadratic polynomial. This can be considered *polynomial regression*. In simple linear regression in Section 5.2.1, we created a matrix $\mathbf{A}_1$ whose columns consisted of the data and a ones vector. Another way to interpret this matrix is $\mathbf{A}_1 = \begin{bmatrix} \mathbf{x}^1 & \mathbf{x}^0 \end{bmatrix}$ powers of our input data. Here, we will use per-capita GDP as the input variable. Unlike Section 5.1, the matrix of powers of x_i will not be square because we will only include $m+1$ columns, where m is the desired degree of the polynomial. To approximate the data by a quadratic, we will use the matrix $\mathbf{A}_2$ given by

$$\mathbf{A}_2 = \begin{bmatrix} 1 & x_0 & x_0^2 \\ 1 & x_1 & x_1^2 \\ \vdots & \vdots & \vdots \\ 1 & x_{n-1} & x_{n-1}^2 \end{bmatrix}.$$

We begin by creating this matrix as a NumPy array. The function `make_power_matrix()` below takes as input a vector and a maximum degree. It returns a matrix whose columns are consecutive powers of the input vector, from degree 0 up to the specified degree[1].

---

[1] This function uses basic NumPy techniques for clarity. Many people use the `PolynomialFeatures` class from `scikit-learn` for generating this type of matrix; an example is given online at la4ds.net/5-2.

```
def make_power_matrix(xdata, degree):
  """stack the powers of the xdata into columns
     of a matrix to use in finding LS solution

  inputs:
    xdata:  column vector of input data
    degree: maximum degree to raise data to

  output:

    power_matrix: matrix whose columns are the powers of xdata from 0 to degree
  """

  # Convert to ndarray, for instance for Pandas Series
  if type(xdata) != np.ndarray:
    xdata = np.array(xdata)

  # If passed a vector, convert to column array
  if len(xdata.shape) == 1:
    xdata = xdata[:, np.newaxis]

  # Initialize the first column
  power_matrix = np.ones_like(xdata)

  # Then consecutively add the powers
  for i in range(1,degree+1):
    power_matrix = np.hstack( (power_matrix, xdata**i) )

  return power_matrix
```

The first five rows of the `A2` matrix are shown below:

```
A2 = make_power_matrix(covid['gdp_norm'], 2)
print(A2[:5])
```

```
[[1.00000000e+00 4.70612673e+01 2.21476288e+03]
 [1.00000000e+00 7.47386695e+01 5.58586872e+03]
 [1.00000000e+00 5.20721990e+01 2.71151391e+03]
 [1.00000000e+00 4.39380424e+01 1.93055157e+03]
 [1.00000000e+00 8.11141428e+01 6.57950417e+03]]
```

Let's confirm that the columns of `A2` are linearly independent by checking that the determinant of its Gram matrix is nonzero:

```
la.det(A2.T @ A2)
```

495814305459.11523

Then the LS solution for the coefficients of the quadratic is given by $\mathbf{A}_2^\dagger \mathbf{y}$, which is calculated as follows:

```
c2 =la.pinv(A2) @ covid['cases_norm']
print(c2)
```

```
[ 2.00857605e+01 -6.91061062e-01  6.39882763e-03]
```

The resulting quadratic is approximately $p(x) = 20.1 - 0.691x + 0.0064x^2$. The resulting fit is shown in Fig. 5.6. Code to generate this plot is available online at la4ds.net/5-2. The quadratic fit is nonmonotonic — it decreases and then increases. Several possible explanations for this are:

- It may be some real phenomenon that causes this in the data.

- It may be because of randomness of the data.

- It may because of limitations with such a low-order fit.

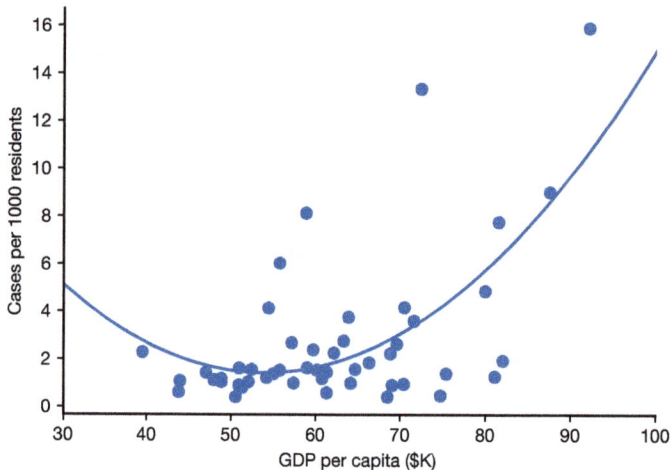

Fig. 5.6: Least-squares quadratic fit to COVID data.

The squared error is easily calculated by substituting this solution back into the equation $\|\mathbf{A}_2\mathbf{c} - \mathbf{y}\|^2$:

```
la.norm(A2 @ c2 - covid['cases_norm'])**2
```

295.2365377484128

This is the lowest squared error that we have achieved yet. Since the overall trend seems from the graph to be that the normalized COVID rates increase with normalized GDPs, let's see if a higher-order polynomial fits the data better. The following code determines the cubic (third-order) least-squares fit. The resulting curve is shown with the data in Fig. 5.7.

```
A3 = make_power_matrix(covid['gdp_norm'], 3)
c3 =la.pinv(A3) @ covid['cases_norm']
print(c3)
```

```
[-6.59326652e+01   3.48324894e+00  -5.92229731e-02   3.34328413e-04]
```

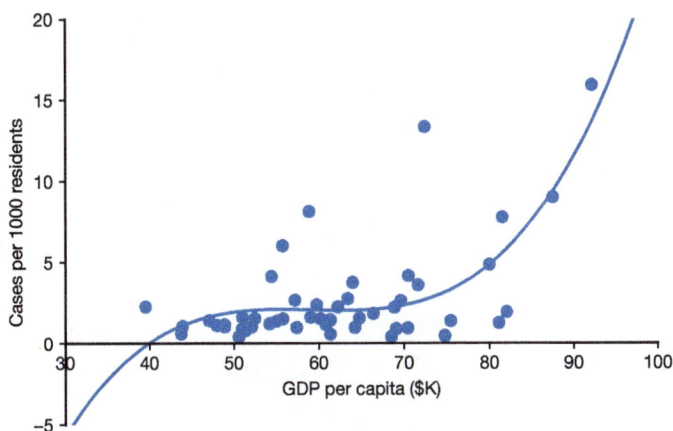

Fig. 5.7: Least-squares cubic fit to COVID data.

The cubic polynomial shows an interesting trend in that it is basically flat for most values of the percentage of GDP per capita, but then increases when the GDP per capita exceeds \$75K. It has the deficiency that it goes negative for small values of the Urban Index. However, those values are outside the range of the data, so this is not necessarily a severe problem. Let's check the error achieved by the cubic polynomial fit:

```
la.norm(A3 @ c3 - covid['cases_norm'])**2
```

```
260.6011960010392
```

The error has been reduced again from 295 for the quadratic fit to 260.6 for the cubic fit. The error will always decrease if we increase the order of the polynomial fit, but the resulting polynomial may not be useful as a model for the data, as we have seen before. The graph below shows the error as a function of the degree of the polynomial for degrees up to 18:

```
errors = []
max_deg=18
for deg in range(1,max_deg):
  Ax = make_power_matrix(covid['gdp_norm'], deg)
  cx =la.pinv(Ax) @ covid['cases_norm']
  ex = la.norm(Ax @ cx - covid['cases_norm'])**2
  errors += [ex]

plt.plot(range(1,max_deg), errors);

plt.xlabel('Degree of polynomial fit')
plt.ylabel('Norm squared of error');
```

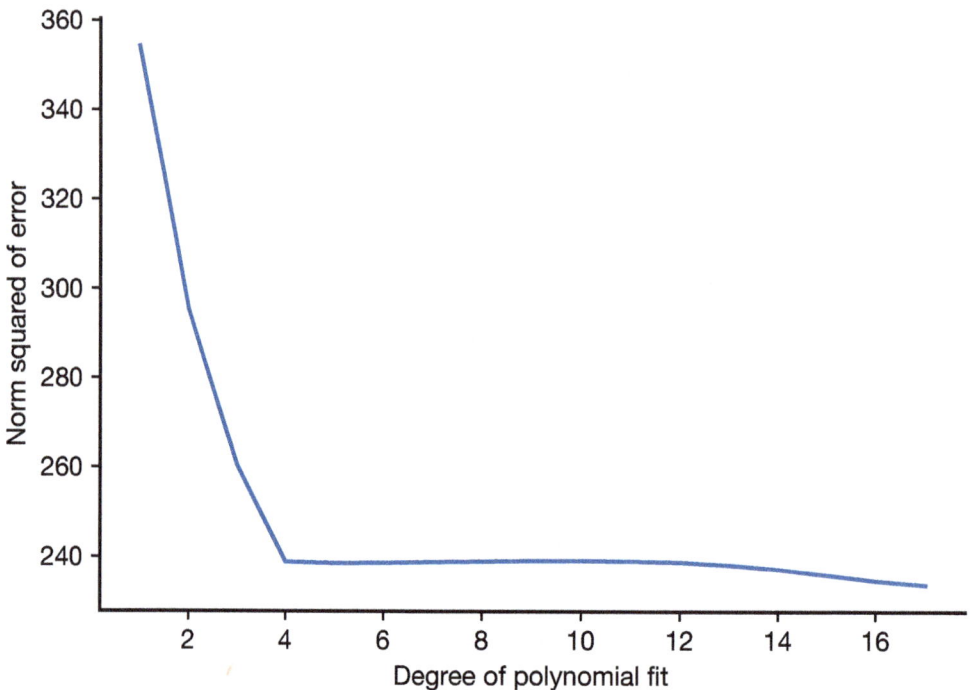

Beyond this point, numerical errors make the polynomial approximation process break down. But we can see that up through degree 18, there is little gain in using a polynomial fit with degree greater than 4. This is typical behavior, and this transition from a steep decrease in error to a more gradual one (an "elbow in the curve") is often used to select the degree of fit.

Terminology review and self-assessment questions

Interactive flashcards to review the terminology introduced in this section and self-assessment questions are available at la4ds.net/5-2, which can also be accessed using this QR code:

5.3 Chapter Summary

This chapter focused on problems of finding polynomials to fit a set of data points. We started with showing how to find a polynomial to exactly fit a set of data points by solving a system of linear equations, provided the degree of the polynomial is sufficiently large. Then we considered a special case where the number of data points is usually much larger than the polynomial degree: linear regression. We showed how to use ordinary least squares to find the solution that minimizes the total squared error from the polynomial to the data. Then we applied the same approach to approximate polynomial fitting/polynomial regression.

Access a list of key take-aways for this chapter, along with interactive flash-cards and quizzes at la4ds.net/5-3, which can also be accessed using this QR code:

6

Transforming Data

This chapter focuses on techniques to represent data in different ways. I will show how to represent a group of vectors using another set of universal or specific vectors, called a *basis*. We will investigate one important application of this, which is transforming time-domain signals into the frequency domain. I will also discuss techniques to represent a group of vectors by a set of vectors with lower dimensionality. We will investigate applications of this to the reception of communication signals in noise and to dimensionality reduction.

6.1 Representing a Vector Using Projections: Spanning Sets and Bases

As shown in Section 2.6.2, vector projection allows us to determine how to represent an arbitrary n-vector $\mathbf{b}$ in terms of a reference vector $\mathbf{a}$. Then given a sufficient set of reference n-vectors $\mathbf{a}_0, \mathbf{a}_1, \ldots$, we should be able to reconstruct the original vector in terms of its vector projections $\text{proj}_{\mathbf{a}_0} \mathbf{b}$, $\text{proj}_{\mathbf{a}_1} \mathbf{b}$, $\ldots$. Let's use an example to demonstrate some conditions in which this is possible.

In Section 3.4.3, we showed how matrix multiplication can be used to rotate a set of data points in two-dimensional space. Let's delve deeper into this idea of representing a vector using a rotated set of axes to build the basic concepts we need for general vector representations. Let's start by formalizing the way we usually represent a vector in two- or three-dimensional Euclidean space. Any vector in these spaces can be written as a sum of scaled versions of unit vectors from the *standard basis*:

DEFINITION

standard basis

In a Euclidean vector space, the *standard basis* consists of a set of unique vectors whose components are all zeros except for a single 1. For the real plane, $\mathbb{R}^2$, the standard basis is

$$\{\mathbf{e}_x = [1,0], \ \mathbf{e}_y = [0,1]\},$$

and for three-dimensional Euclidean space, $\mathbb{R}^3$, the standard basis is

$$\{\mathbf{e}_x = [1,0,0], \ \mathbf{e}_y = [0,1,0], \ \mathbf{e}_z = [0,0,1]\}.$$

It is worth noting two important properties of the vectors in any standard basis. Below, we show the properties for $\mathbb{R}^3$ as an example.

DOI: 10.1201/9781032664088-6

1. They are normal vectors: $\|\mathbf{e}_x\| = \|\mathbf{e}_x\| = \|\mathbf{e}_z\| = 1$

2. They are mutually orthogonal, meaning that any pair of vectors is orthogonal. In $\mathbb{R}^3$, $\mathbf{e}_x \cdot \mathbf{e}_y = 0$, $\mathbf{e}_x \cdot \mathbf{e}_z = 0$, and $\mathbf{e}_y \cdot \mathbf{e}_z = 0$.

Any set of vectors satisfying these two properties is called an *orthonormal set*:

DEFINITION

orthonormal set (of vectors)

A set of vectors $\{\mathbf{a}_0, \mathbf{a}_1, \ldots, \mathbf{a}_{n-1}\}$ is an *orthonormal set* if and only if:

1. The vectors are all normal; i.e., $\|\mathbf{a}_i\| = 1$ for all $i = 0, 1, \ldots, n-1$.

2. The vectors are mutually orthogonal; i.e., $\mathbf{a}_i \cdot \mathbf{a}_j$ for all $i \neq j$ in $0, 1, \ldots, n-1$.

We can interpret the components of an arbitrary vector as the coefficients of a linear combination of the standard basis vectors. For example, if $\mathbf{w} = [a, b]$, then $\mathbf{w} = a\mathbf{e}_x + b\mathbf{e}_y$.

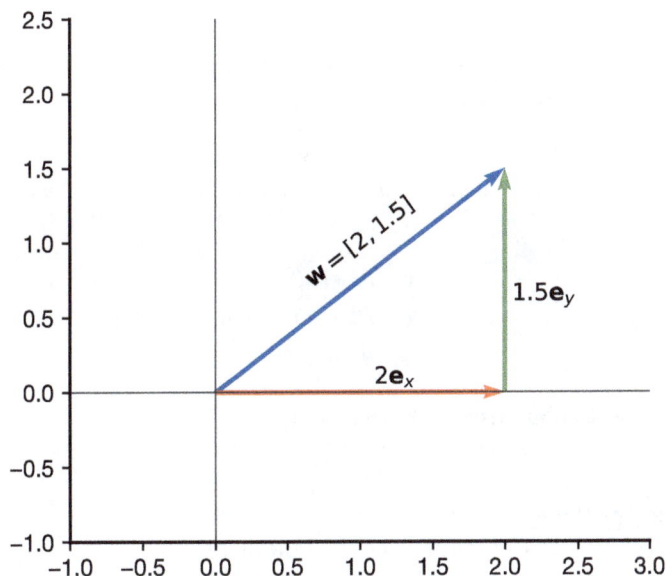

Fig. 6.1: Representation of $\mathbf{w} = [2, 1.5]$ as linear combination of standard basis vectors $\mathbf{e}_x$ and $\mathbf{e}_y$.

Example 6.1: Representation of Vector Using Standard Basis Vectors

Consider the vector $\mathbf{w} = [2, 1.5]$. Fig. 6.1 shows our usual plot of $\mathbf{w}$ along with two vectors showing $\mathbf{w} = 2\mathbf{e}_x + 1.5\mathbf{e}_y$. Note also that the vectors $2\mathbf{e}_x$ and $1.5\mathbf{e}_y$ are the vector projections of $\mathbf{w}$ onto $\mathbf{e}_x$ and $\mathbf{e}_y$:

```
import numpy as np
from numpy.linalg import norm

w = np.array([2, 1.5])
ex = np.array([1, 0])
ey = np.array([0, 1])

wpx = w@ex * ex / norm(ex)**2
print(f'Projection of z onto e_x = {wpx}')

wpy = w@ey * ey / norm(ey)**2
print(f'Projection of z onto e_y = {wpy}')
```

```
Projection of z onto e_x = [2. 0.]
Projection of z onto e_y = [0. 1.5]
```

Suppose we create new axes represented by unit vectors $\mathbf{e}_{x,\theta}$ and $\mathbf{e}_{y,\theta}$ that are rotated θ degrees from the standard basis.

Example 6.2: Standard Basis Vectors Rotated 60° Counter-Clockwise

If we rotate the standard unit vectors by 60° counter-clockwise, the resulting unit vectors are shown in Fig. 6.2. From trigonometry, we can see that the unit vectors for the axes are

$$\mathbf{e}_{x,\theta} = [\quad \cos\theta \quad \sin\theta]$$
$$\mathbf{e}_{y,\theta} = [\quad \cos(\theta+90°) \quad \sin(\theta+90°)]$$
$$= [-\sin\theta \quad \cos\theta].$$

Then for the particular case of 60°, we define:

```
theta_r = np.deg2rad(60)
ex60 = np.array([ np.cos(theta_r), np.sin(theta_r)])
ey60 = np.array([ -np.sin(theta_r), np.cos(theta_r)])

print(f'x-axis rotated by 60 deg CCW = {ex60}')
print(f'y-axis rotated by 60 deg CCW = {ey60}')
```

```
x-axis rotated by 60 deg CCW = [0.5        0.8660254]
y-axis rotated by 60 deg CCW = [-0.8660254  0.5       ]
```

Note that rotating the axes does not change the fact that they are an orthonormal set. First, let's check that they are still normal vectors:

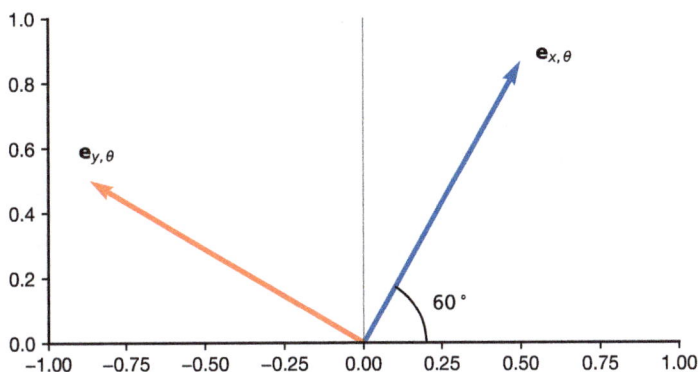

Fig. 6.2: Vectors from rotating standard unit vectors by 60° counter-clockwise.

```
norm(ex60), norm(ey60)
```

```
(1.0, 1.0)
```

Now let's check that they are still orthogonal:

```
np.round(ex60 @ ey60, 10)
```

```
-0.0
```

Now lets consider what happens if we transform a vector by projecting it onto a rotated set of axes, $\mathbf{e}_{x,\theta}$ and $\mathbf{e}_{y,\theta}$.

Example 6.3: Projection of Vector onto Axes Rotated 60° CCW

The projection of the vector $\mathbf{w} = [2, 1.5]$ onto $\mathbf{e}_{x,60°}$ and $\mathbf{e}_{y,60°}$ is

```
wpx = w @ ex60
wpy = w @ ey60

print(f'The scalar projections of w onto the rotated axes: {wpx:.2f}, {wpy:.2f}')
```

```
The scalar projections of w onto the rotated axes: 2.30, -0.98
```

Now let's visualize the vector projections onto these axes. The rotated axis vectors are shown as thin, black, non-transparent vectors, and the projections are shown as thicker, colored, and partially transparent vectors:

```
wx = wpx * xp
wy = wpy * yp
plotvec(xp, yp, width=0.005, colors=['k', 'k'])
plotvec(wx, wy, newfig=False, width=0.02, color_offset=3, alpha=0.7)
plt.xlim([-2, 2])
plt.ylim([-1, 2]);
```

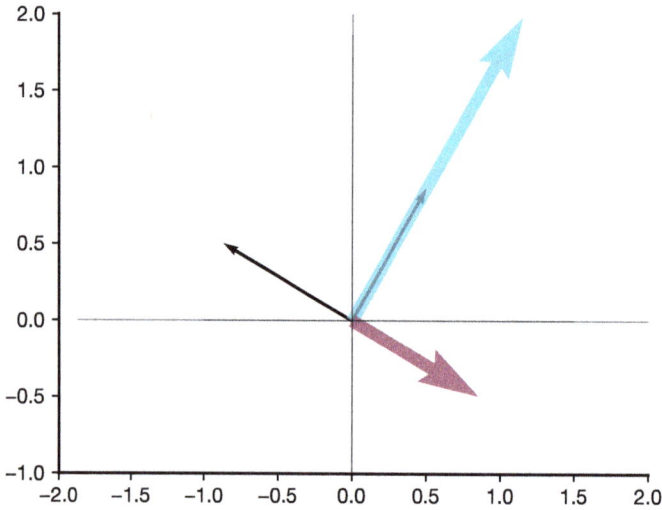

Finally, let's make a figure that shows the representations before and after rotation. The figure below shows:

1. The original vector **w** as a thin non-transparent vector.

2. The linear combination of the vector projections of **w** onto the rotated axes. This second vector is shown as a thicker, semi-transparent arrow.

   The result shows that the linear combination of the vector projections onto the rotated axes completely reconstructs the original vector **w**.

```
plotvec(w, width=0.005, colors=['k'])
plotvec(wx + wy, newfig=False, width=0.02, color_offset=3, alpha=0.7)
```

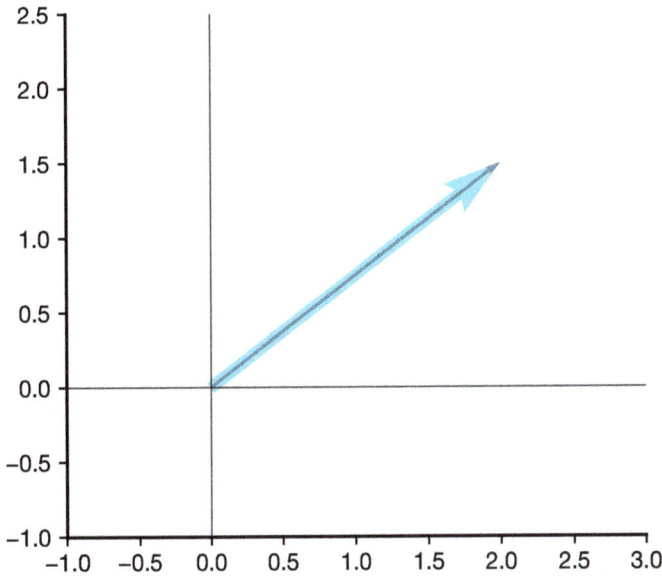

We can confirm this by checking the norm of the error vector, which is the difference between the original vector and the sum of the vector projections:

```
e = w - (wx + wy)
print(f'The energy in the error vector is {norm(e) : .3f}')
```

```
The energy in the error vector is 0.000
```

In the results above, we still show **w** and the sum of the vector projections **on the original axes**. To translate **w** to the rotated axes, we create a new vector $\mathbf{w}_\theta$ for which the components are the scalar projections of **w** onto the rotated axes:

Example 6.4: Projection onto Rotated Axes

Let $\mathbf{w}_{60}$ be the vector whose components are the scalar projections of **w** onto $\mathbf{e}_{x,60°}$ and $\mathbf{e}_{y,60°}$:

```
w60 = np.array([wpx, wpy])
```

If we plot $\mathbf{w}_{60}$ and **w** on the same axes, then $\mathbf{w}_{60}$ is equivalent to a 60° **clockwise** rotation of **w**. (In other words, when we project onto a basis that is rotated 60° counter-clockwise from the standard basis, the resulting vector representation is equivalent to a 60° clockwise rotation of the original vector.)

```
plotvec(w, w60, labels=['$\mathbf{w}$', '$\mathbf{w}_{60}$'])
plt.xlim([-1, 3])
plt.ylim([-1.5, 2])
```

(-1.5, 2.0)

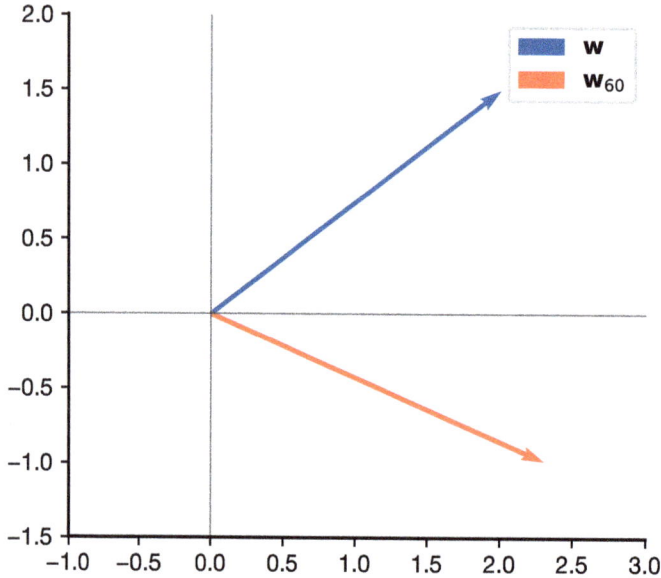

We can confirm this using our formula for the angle between two vectors:

```
theta = np.arccos(w60 @ w / norm(w) / norm(w60))
print(f'theta = {np.rad2deg(theta): .2g}')
```

```
theta =  60
```

The representation of **w** using the rotated axes has the same length as the original vector **w**:

```
print(f'Length of w = {norm(w) : .2f}')
print(f'Length of w projected onto rotated axes = {norm(w60) : .2f}')
```

```
Length of w = 2.50
Length of w projected onto rotated axes = 2.50
```

Any vector in $\mathbb{R}^2$ can be represented using these new axes – we can simply rotate a vector by $60°$ clockwise to find its representation on the rotated axes.

We say that $\mathcal{S} = \{\mathbf{e}_{x,\theta}, \mathbf{e}_{y,\theta}\}$ is a *spanning set* for $\mathbb{R}^2$ (or say that $\mathcal{S}$ *spans* $\mathbb{R}^2$):

DEFINITION

spanning set (vector space)

A set of vectors $\mathcal{S} = \{\mathbf{x}_0, \mathbf{x}_1, \ldots, \mathbf{x}_{n-1}\}$ is a *spanning set* for a vector space $\mathbb{V}$ if every vector in $\mathbb{V}$ can be represented as a linear combination of the vectors in S.

Note that we cannot remove either of the vectors from $\mathcal{S}$ and still be able to represent everything in $\mathbb{R}^2$. For example, we need both rotated axis vectors to represent $\mathbf{w}$. We say that $\mathcal{S}$ is *minimal*:

DEFINITION

minimal (spanning set)

A spanning set $\mathcal{S}$ for a vector space $\mathbb{V}$ is *minimal* if the removal of any member of $\mathcal{S}$ would stop it from being a spanning set for $\mathbb{V}$.

An equivalent condition for a spanning set to be minimal is that the vectors in the spanning set are linearly independent.

A minimal spanning set is also called a basis:

DEFINITION

basis (vector space)

A set of vectors $\mathcal{S}$ is a basis for a vector space $\mathbb{V}$ if $\mathcal{S}$ is a minimal spanning set for $\mathbb{V}$.

The plural of basis is *bases* (pronounced "base-ease").

We can always find a basis that is an orthonormal set, and such a basis is called an orthonormal basis:

DEFINITION

orthonormal basis (for set of vectors)

A set of vectors $\mathcal{S}$ is an orthonormal basis for a set of vectors $\mathcal{V}$ if

1. $\mathcal{S}$ is an orthonormal set, and

2. $\mathcal{S}$ is a minimal spanning set for $\mathcal{V}$.

There are generally multiple bases for any vector space $\mathbb{V}$, but they all have the same *cardinality*:

DEFINITION

cardinality (set)

The *cardinality* of a set $\mathcal{S}$ is denoted $|\mathcal{S}|$ and is equal to the number of elements in the set.

We will only consider the cardinality for finite sets. The cardinality of a basis for a vector space $\mathbb{V}$ is called its *dimension*:

DEFINITION

dimension (vector space)

The dimension of a vector space $\mathbb{V}$, denoted $\dim \mathbb{V}$, is the cardinality of a basis for $\mathbb{V}$.

For $\mathbf{R}^n$, one basis is $\mathscr{B}_n = \{\mathbf{e}_0, \mathbf{e}_1, \ldots, \mathbf{e}_{n-1}\}$, where each $\mathbf{e}_i$ is taken to be the n-vector that consists of all zeros except for a 1 in position i. Thus, the dimension of $\mathbf{R}^n$ is $|\mathscr{B}_n| = n$. Even when limiting to real n-vectors, there are infinitely many other vector spaces other than $\mathbb{R}^n$. One common way to form such a vector space is to specify a set of vectors $\mathcal{V}$ and then specify the vector space $\mathbb{V}$ as the *span* of $\mathcal{V}$:

DEFINITION

span (set of vectors)

Given a set of vectors $\mathcal{V}$, the *span* of $\mathcal{V}$, denoted $\mathrm{span}(\mathcal{V})$ is the vector space that consists of all linear combinations of the vectors in $\mathcal{V}$.

I.e., if $|\mathcal{V}| = n$, then $\mathrm{span}(\mathcal{V})$ contains every vector of the form $c_0\mathbf{v}_0 + c_1\mathbf{v}_1 + \ldots + c_{n-1}\mathbf{v}_{n-1}$ for all $v_i \in \mathcal{V}$ and all real constants c_i. By definition, $\mathcal{V}$ is a spanning set for $\mathrm{span}(\mathcal{V})$. If $\mathcal{V}$ is a set of linearly independent vectors, then $\mathcal{V}$ is a basis for $\mathrm{span}(\mathcal{V})$.

Example 6.5: Another Basis for $\mathbb{R}^3$

Consider the set $\mathcal{V} = \left\{[1,1,0]^\mathsf{T}, [1,0,1]^\mathsf{T}, [0,1,1]^\mathsf{T}\right\}$. These vectors are linearly independent, and thus $\mathcal{V}$ is a basis for $\mathrm{span}(\mathcal{V})$. Note that $\dim \mathrm{span}(\mathcal{V}) = 3$, $\mathrm{span}(\mathcal{V}) \subset \mathbb{R}^3$, and $\dim \mathbb{R}^3 = 3$. It can be shown that $\mathcal{V}$ is a basis for $\mathbb{R}^3 = 3$.

Example 6.6: Example of a Vector Space of 3-Vectors Other Than $\mathbb{R}^3$

Consider the set $\mathcal{W} = \left\{ [1, -1, 0]^\mathsf{T}, [1, 0, -1]^\mathsf{T} \right\}$. These vectors are linearly independent, and thus $\mathcal{W}$ is a basis for span$(\mathcal{W})$. Since $|\mathcal{W}| = 2$, span$(\mathcal{W})$ is a vector space of dimension 2 and thus cannot be equal to $\mathbb{R}^3$.

It is also easy to see that there are vectors in $\mathbb{R}^3$ that are not in span$(\mathcal{W})$, such as $[1, 0, 0]^\mathsf{T}$, $[0, 1, 0]^\mathsf{T}$, and $[0, 0, 1]^\mathsf{T}$. In fact, it can be shown that if we interpret the vectors as points in three-dimensional space, then span$(\mathcal{W})$ is the plane defined by $\{(x, y, z) \mid x + y + z = 0\}$, which is shown in Fig. 6.3. An interactive version of this plot is available at la4ds.net/6-1.

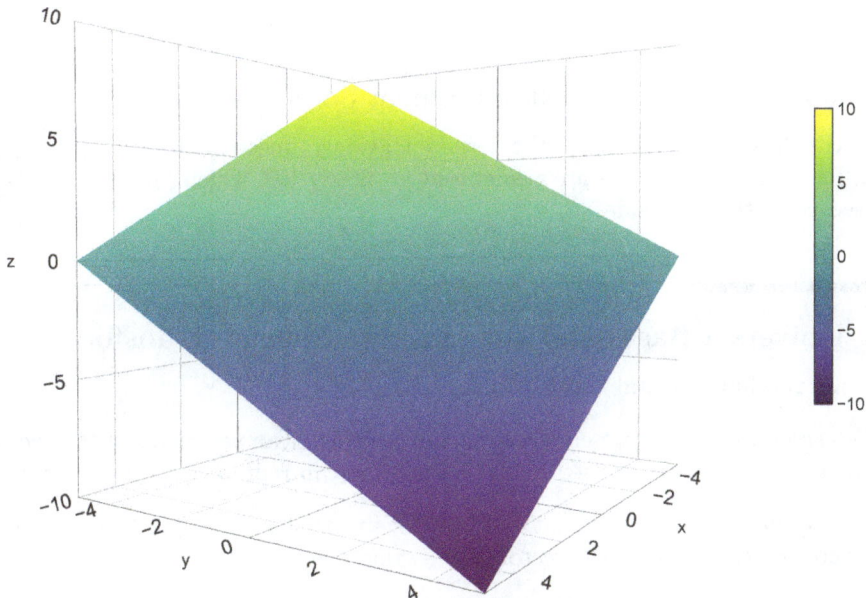

Fig. 6.3: The three dimensional vector space with basis $\{[1, -1, 0]^\mathsf{T}, [1, 0, -1]^\mathsf{T}\}$ can be viewed as the plane shown, which corresponds to the solution set of $x + y + z = 0$.

6.1.1 Applications of Alternative Bases

Alternative bases are useful for many different purposes in data science and engineering:

1. Alternative bases can be used to provide an alternative interpretation of data in terms of features such as frequency.

2. Alternative bases can be used when we want to detect signals embedded in noise by projecting the noisy signal into those dimensions that enhance the signal components and suppress the noise.

3. Alternative bases can be used when we want to reduce the dimensionality of a set of vectors. Instead of throwing away some particular elements of a vector, we can find a basis that captures the most important features of a set of vectors, resulting in a smaller number of elements in a lossy representation. This can be useful for:

   1. Plotting high-dimensional data by projecting the data onto a basis that reduces the data to 3 dimensions or fewer.

   2. Applying two-dimensional statistical techniques (like 2-D regression) to high-dimensional data.

   3. Reducing the complexity of statistical methods, signal processing techniques, or machine learning algorithms by eliminating redundant or irrelevant information.

   4. Performing data compression by preserving the most important information and discarding the least important information.

I will demonstrate these in the following sections. In the next section, I show how an alternative basis constructed from complex sinusoids can be used to find the frequency content of time-domain vectors.

Terminology review and self-assessment questions

Interactive flashcards to review the terminology introduced in this section and self-assessment questions are available at la4ds.net/6-1, which can also be accessed using this QR code:

6.2 Universal Bases and the Discrete Fourier Transform

There are two fundamental types of bases for sets of vectors:

1. *Universal bases* can represent every vector in a Euclidean vector space, $\mathbb{R}^n$. For example, the rotated axis vectors used in the last section can represent every vector in $\mathbb{R}^2$.

2. *Set-specific bases* can represent every vector in the span of a specified set but generally cannot represent every real vector of the same size.

Why do we not always use a universal basis?

Recall that the cardinality of a basis is the dimensionality of the vector space that it can represent. The dimensionality of the Euclidean vector space $\mathbb{R}^n$ is n, and so n basis vectors are required. However, suppose we have a set $\mathcal{S}$ of n-vectors, where the cardinality is $m < n$. Then the dimensionality of span$(\mathcal{S})$ is at most m, and thus fewer than n basis vectors are required. If m is much smaller than n, this can make a significant difference in computation. In Example 6.14, I show how using a smaller basis can help distinguish a signal from noise in a digital communication system.

In this section, we consider a particular universal basis made out of sinusoids. This type of basis is often applied to time-series data. When time-series data is represented using sinusoidal basis functions, the representation characterizes the frequency content of the signal. This frequency representation is useful in many applications, including:

1. **Audio:** the frequency representation can be used to analyze or manipulate the frequency content of speech or music signals. For instance, the frequency representation can be used for equalization, pitch shifting, or noise removal.

2. **Financial data:** the frequency representation can be used to identify patterns in price changes of stocks or commodities.

3. **Mechanical signals:** time-series data from mechanical systems, such as an engine, turbine, or machine, can be used to identify vibrations.

4. **Biological data:** time-series data, such as electrical signals from the brain or heart, can be analyzed to detect medical problems.

6.2.1 Sinusoidal Basis

Let's begin by generating a set of sinusoidal basis vectors. We can generate such a set as the rows of a special matrix called a Discrete Fourier Transform (DFT) matrix. We will show how to generate a DFT matrix using NumPy and how to interpret it using plots. Readers who want to understand the math behind DFT matrices can refer to the Wikipedia page: https://en.wikipedia.org/wiki/DFT_matrix.

The DFT matrix for length-64 vectors can be created using NumPy functions[1] as follows:

```python
import numpy as np

dft_len = 64
dft = np.fft.fft(np.eye(dft_len), norm='ortho')
print(np.round(dft, 4))
```

```
[[0.125 +0.j      0.125 +0.j       0.125 +0.j      ... 0.125 +0.j
   0.125 +0.j      0.125 +0.j      ]
 [0.125 +0.j      0.1244-0.0123j 0.1226-0.0244j ... 0.1196+0.0363j
   0.1226+0.0244j 0.1244+0.0123j]
 [0.125 +0.j      0.1226-0.0244j 0.1155-0.0478j ... 0.1039+0.0694j
   0.1155+0.0478j 0.1226+0.0244j]
 ...
 [0.125 +0.j      0.1196+0.0363j 0.1039+0.0694j ... 0.0793-0.0966j
   0.1039-0.0694j 0.1196-0.0363j]
 [0.125 +0.j      0.1226+0.0244j 0.1155+0.0478j ... 0.1039-0.0694j
   0.1155-0.0478j 0.1226-0.0244j]
 [0.125 +0.j      0.1244+0.0123j 0.1226+0.0244j ... 0.1196-0.0363j
   0.1226-0.0244j 0.1244-0.0123j]]
```

---

[1]The equivalent function in PyTorch is `torch.fft.fft()`.

A few comments:

- The function name `fft` stands for *fast Fourier transform*, which is a fast technique for computing the DFT when the size is a power of 2. The term FFT is commonly used to refer to the DFT and `np.fft.fft()` works even when the length is not a power of 2.

- There are two instances of "fft" in the function call because the `fft()` function is part of NumPy's `fft` module.

- The `norm='ortho'` keyword argument is to make these sinusoids have unit norm. The standard FFT does not give vectors with unit norm and instead requires different normalization in the inverse FFT (IFFT) function.

- Most entries in the DFT matrix consist of the sum or difference of two components, one of which has a "j" at the end. This is because these DFT entries are complex numbers. Python uses the suffix j on a numeric value to indicate an imaginary number. (The letter j is typically used for this purpose by electrical engineers because i is used to denote current in a circuit.)

We can interpret the complex entries of a given row of the DFT matrix as follows:

- The *real* parts are samples of a cosine function at some frequency.

- The *imaginary* parts are samples of the negative of a sine function at the same frequency.

The frequencies increase with the row number for the first half of the rows, and then decrease during the second half of the rows. Examples of these sinusoids are shown in Fig. 6.4, which includes separate plots of the real and imaginary components of rows 0, 1, and 4 of the FFT matrix[2].

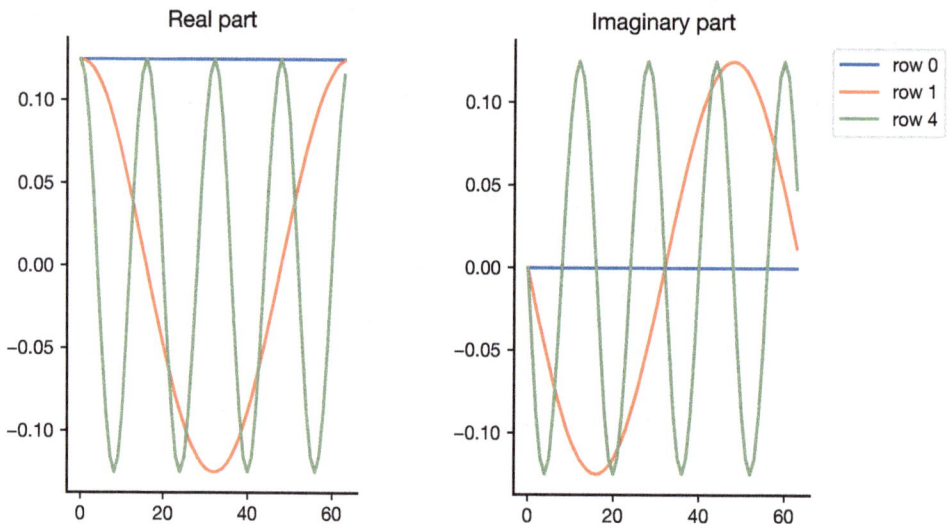

Fig. 6.4: Plots of the real and imaginary components of FFT vectors 0, 1, and 4.

---

[2]Code to create this plot is available at la4ds.net/6-2

Let's confirm that all of the rows have unit norm:

```
np.linalg.norm(dft, axis=1)
```

```
array([1., 1., 1., 1., 1., 1., 1., 1., 1., 1., 1., 1., 1., 1., 1., 1.,
       1., 1., 1., 1., 1., 1., 1., 1., 1., 1., 1., 1., 1., 1., 1., 1.,
       1., 1., 1., 1., 1., 1., 1., 1., 1., 1., 1., 1., 1., 1., 1., 1.,
       1., 1., 1., 1., 1., 1., 1., 1., 1., 1., 1., 1., 1., 1.])
```

The frequencies are such that all of the sinusoids at all of the frequencies are orthogonal, and this results in the complex vectors being orthogonal. For complex vectors **a** and **b**, the inner product is defined as $\mathbf{a} \cdot \mathbf{b}^*$, where the elements of $\mathbf{b}^*$ are the complex conjugates of the elements in **b**. Then if we take any two rows, we will get an inner product of zero:

```
np.round(dft[0]@np.conj(dft[4]), 10), np.round( dft[1]@np.conj(dft[4]), 10)
```

```
((-0+0j), 0j)
```

Let's start by analyzing the frequency content of a simple signal before using the DFT to analyze real data.

Example 6.7: DFT of Square Wave

We can generate a square wave signal using `scipy.signal.square()`. Below I generate a length-64 square wave with four cycles (I chose the parameters to get exactly four cycles, starting with half a cycle):

```
import scipy.signal
offset=4
t = np.linspace(0, 8*(64+offset)/64*np.pi, 64+offset)[offset:]
sq = scipy.signal.square(t)

plt.plot(sq);
plt.title('Length-64 square wave with 4 cycles');
```

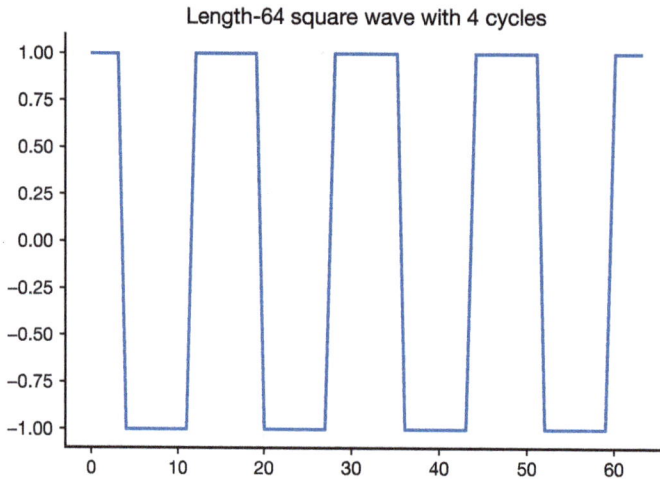

Length-64 square wave with 4 cycles

We can determine the frequency content of this wave by multiplying its vector representation by the rows of the DFT matrix. Let's do a few frequency components individually before automating the process. Row 0 is a DC signal, so it determines any DC component in the signal:

```
sq @ dft[0]
```

```
0j
```

Because we want to look at the scalar projection with many different rows, let's make a function to print out the scalar projections given a list of rows. If the result is very small, let's provide an option to suppress the output:

```
def print_projections(signal, dft_matrix, rows, suppression_threshold = 0):
    ''' print the scalar projection between the signal vector
        and the rows of the dft matrix    '''
    for row in rows:
        proj = sq @ dft[row]
        if abs(proj) > suppression_threshold :
            print(f'scalar projection of row {row}: {proj:.2f}')
```

Here are the values for rows 1 and 2:

```
print_projections(sq, dft, [1, 2])
```

```
scalar projection of row 1: 0.00-0.00j
scalar projection of row 2: 0.00+0.00j
```

Interestingly, all of these scalar projections are 0. This is because the waveform does not contain any of those low frequencies. Now consider the scalar projection with rows 3 and 4:

```
print_projections(sq, dft, [3, 4], suppression_threshold = 0.05)
```

```
scalar projection of row 4: 5.03+1.00j
```

Although the scalar projection with row 3 is zero, the scalar projection with row 4 is not. For real data, each row k in the DFT has a "partner" at $64-k$ such that if we take twice the real part of the scalar projection onto row k, we get the combined effect of these two rows. The result is a sinusoid that could be considered a very crude approximation of the square wave:

```
plt.plot(sq)
plt.plot(2*np.real(dft[4]@sq * dft[4]));
```

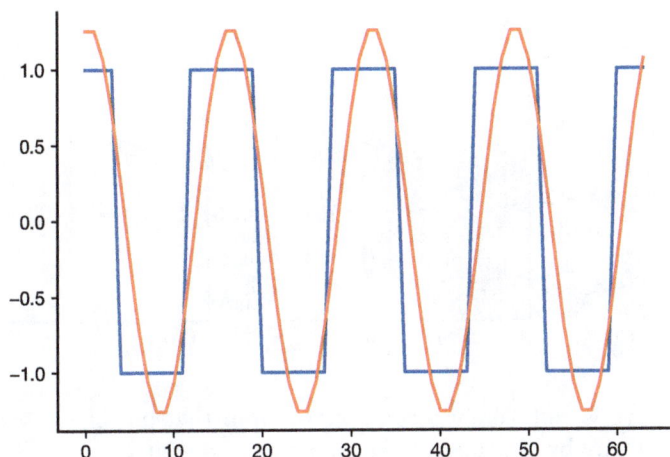

Now let's find what other projections are nonzero for the first half of the DFT functions:

```
print_projections(sq, dft, range(5, 32), suppression_threshold=0.05)
```

```
scalar projection of row 12: -1.50-1.00j
scalar projection of row 20: 0.67+1.00j
scalar projection of row 28: -0.20-1.00j
```

Every 8th frequency component has a nonzero frequency component. The code below combines different numbers of projections. The approximation is plotted versus the original square wave after each new vector projection is added:

```
fig, axs = plt.subplots(1, 3, figsize=(8,4) )

approximation = 2 * np.real(np.conj(dft[4]) @ sq * dft[4])

rows=[4]

for i, row in enumerate([12,20, 28]):
  approximation += 2 * np.real(np.conj(dft[row]) @ sq * dft[row])

  ax = axs[i]
  ax.plot(sq)
  ax.plot(approximation)
  rows += [row]
  ax.set_title(f'Square wave & approximation\nusing rows {rows}')

plt.tight_layout()
```

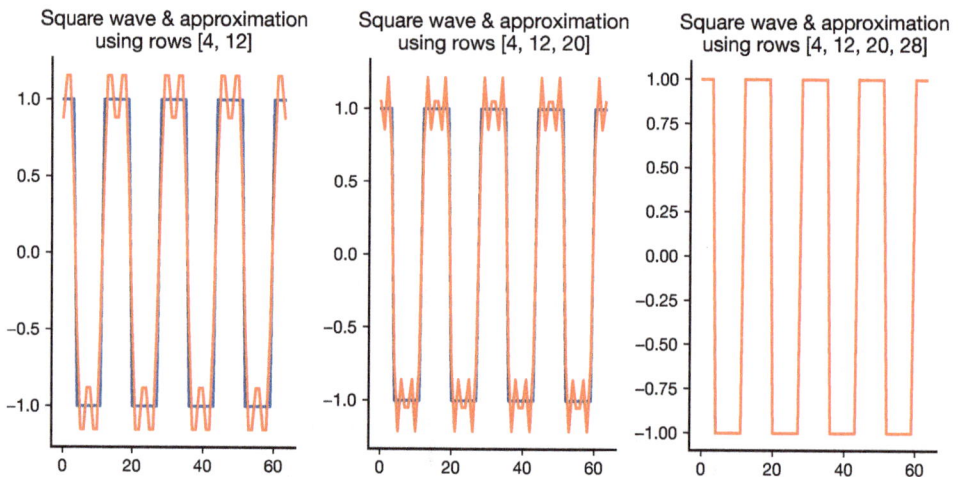

The original square wave is completely reconstructed from the sum of the vector projections, as we can verify by checking the norm of the error signal:

```
np.round(np.linalg.norm(sq - approximation), 10)
```

```
0.0
```

The scalar projections can all be found simultaneously using matrix multiplication. Let **v** be an n-vector, and let $\mathbf{F}_n$ be the $n \times n$ DFT matrix. Then we can get the DFT of **v** is

$$\mathbf{V} = \mathbf{F}_n \mathbf{v}. \tag{6.1}$$

Note that we use an uppercase **V** for the frequency domain representation, and a lowercase **v** for the time domain signal. Equation 6.1 is called the *analysis equation* of the DFT:

DEFINITION

analysis equation

Let $\mathbf{v}$ be an n-vector from some set of vectors $\mathcal{V}$, and let B be a $m \times n$ matrix whose rows are the basis for $\mathcal{V}$. Then the DFT of $\mathbf{v}$ is given by the *analysis equation*,

$$\mathbf{V} = \mathbf{B}\mathbf{v}.$$

The analysis equation may be written in other forms, depending on the application, but always corresponds to projecting the vector that is to be represented onto each of the basis vectors.

The DFT can be found even more efficiently using the Fast Fourier Transform (FFT) when the length of the vector is a power of 2. We can get the DFT projections directly using NumPy using the `np.fft.fft()` function as follows:

```
np.round(np.fft.fft(sq, norm='ortho'), 5)
```

```
array([ 0.      +0.j, 0.      +0.j, 0.      +0.j, 0.      +0.j,
        5.02734+1.j, 0.      +0.j, 0.      +0.j, 0.      +0.j,
        0.      +0.j, 0.      +0.j, 0.      +0.j, 0.      +0.j,
       -1.49661-1.j, 0.      +0.j, 0.      +0.j, 0.      +0.j,
        0.      +0.j, 0.      +0.j, 0.      +0.j, 0.      +0.j,
        0.66818+1.j, 0.      +0.j, 0.      +0.j, 0.      +0.j,
        0.      +0.j, 0.      +0.j, 0.      +0.j, 0.      +0.j,
       -0.19891-1.j, 0.      +0.j, 0.      +0.j, 0.      +0.j,
        0.      +0.j, 0.      +0.j, 0.      +0.j, 0.      +0.j,
       -0.19891+1.j, 0.      +0.j, 0.      +0.j, 0.      +0.j,
        0.      +0.j, 0.      +0.j, 0.      +0.j, 0.      +0.j,
        0.66818-1.j, 0.      +0.j, 0.      +0.j, 0.      +0.j,
        0.      +0.j, 0.      +0.j, 0.      +0.j, 0.      +0.j,
       -1.49661+1.j, 0.      +0.j, 0.      +0.j, 0.      +0.j,
        0.      +0.j, 0.      +0.j, 0.      +0.j, 0.      +0.j,
        5.02734-1.j, 0.      +0.j, 0.      +0.j, 0.      +0.j])
```

I will demonstrate how to use this function to analyze the frequency content in a real biological signal in the following example.

Example 6.8: Analyzing Frequency Content of ECG Data

An electrocardiogram (ECG) is a record of the heart's electrical activity. The file fdsp.net/data/ecg.csv contains the ECG data for a healthy adult. ECGs often contain data for multiple channels, but this contains data only from a single lead (the I lead). Let's start by loading this data and looking at the head of the dataframe:

```
import pandas as pd

ecg_df = pd.read_csv('https://fdsp.net/data/ecg.csv', header=None, skiprows=2)
ecg_df.rename(columns={0 : 'ecg'}, inplace=True)
ecg_df.head()
```

```
        ecg
0   -41.920
1   -44.644
2   -47.030
3   -49.050
4   -50.681
```

This is supposed to be time-series data, but the time stamps are missing. However, the first line of this file says: `Sample Rate,512.242 hertz`, which means that 512.242 samples were taken per second. Thus the time between samples (in seconds) is

```
1 / 512.242
```

```
0.0019522022793913817
```

We can assign a time to each ECG sample and plot versus time as shown below:

```
ecg_df = ecg_df.assign(time = lambda x: x.index / 512.242)

plt.plot(ecg_df['time'], ecg_df['ecg'])
plt.xlabel('time (s)')
plt.ylabel('ECG signal strength ($\mu V$)');
```

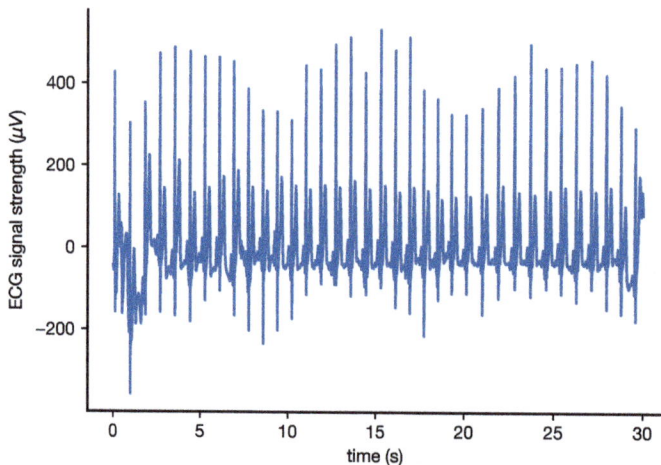

There are some amplitude variations that were probably caused by movement (these ECG results are from a smart watch). Let's zoom in on a few cycles to get a better idea of what the ECG data looks like:

```
s_min = 1370
s_max = 2665
plt.plot(ecg_df['time'].iloc[s_min:s_max], ecg_df['ecg'].iloc[s_min:s_max])
plt.xlabel('time (s)')
plt.ylabel('ECG signal strength ($\mu V$)');
```

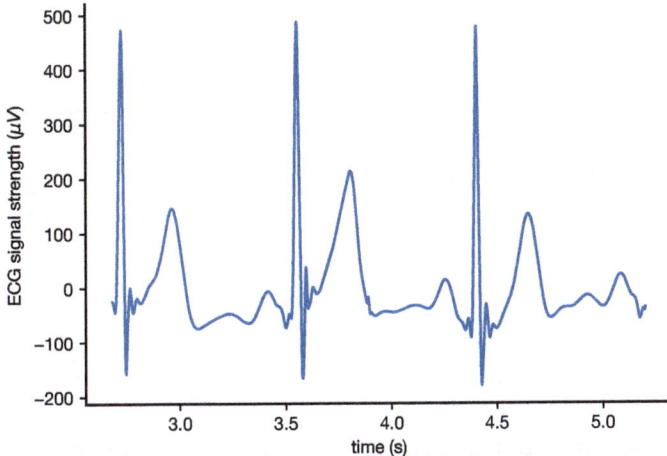

From this plot, we can clearly see the repeating structure of the ECG data. Each cycle of the ECG signal lasts approximately 0.8s (from a visual inspection), implying a heart rate in beats per minute of approximately:

```
60/0.8
```

```
75.0
```

If we want to estimate the heart rate across this data frame, we have a couple of options. We could try to write a function to 1) find the peaks, 2) find the differences between adjacent peaks, and 3) average those differences. However, the time-varying nature of the peaks may make this challenging. As an alternative, we can transform the data into the frequency domain as follows:

```
ecg = ecg_df['ecg']
ECG = np.fft.fft(ecg)
```

Again, I have used capital letters to indicate that `ECG` is a frequency-domain representation of `ecg`. `ECG` is complex, but if we plot its magnitude-squared, the result is proportional to the power at each frequency. The power at each frequency is called the *power spectral density*:

DEFINITION

power spectral density (of a vector)

For a n-vector $\mathbf{x}$ with DFT $\mathbf{X}$, the power spectral density is the power at each frequency component and is given by

$$P[k] = \frac{1}{n}\left\|\mathbf{X}[n]\right\|^2.$$

In the plots and analysis that follow, we are only interested in the relative power at different frequencies, so I am going to omit the normalization term $1/n$. Let's start by plotting the power spectral density for the whole `ECG` vector.

```
plt.plot(abs(ECG)**2);
plt.title('Magnitude-squared of DFT of ECG data');
```

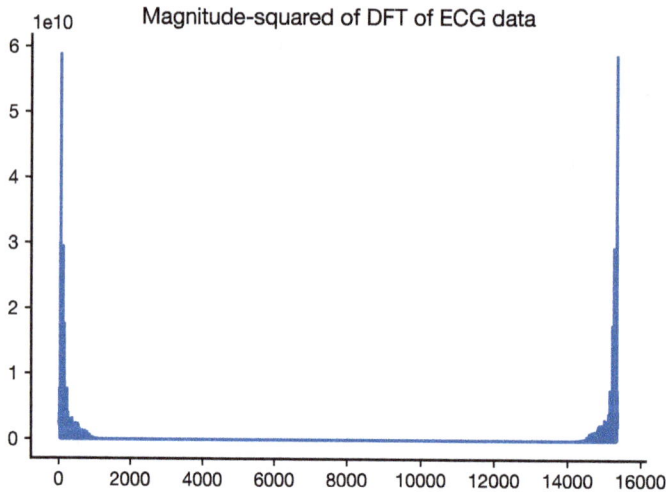

First, note the symmetry in the PSD. As before, only the first half of the DFT data is needed to capture all of the frequency information for a real vector.

Second, two issues make this plot hard to use:

1. The dependent variable in this plot is the index of the rows of the DFT matrix, not frequency.

2. The larger PSD values are concentrated at the lower frequency indices, and we can't really see what is going on when plotting the full range.

To plot against frequency, we can use a helper function called `np.fft.fftfreq()` to return a vector of frequencies. Its arguments are the length of the DFT vectors and the time separation between samples. Since we have 512.242 samples/second, the time separation is the inverse of that value. We can create a vector of frequencies and plot the first few hundred components of the DFT versus frequency as follows:

```
f = np.fft.fftfreq(len(ecg), 1/512.242)

plt.plot(f[:400], 1/len(ECG) * np.abs(ECG[:400])**2)
plt.title('Power spectral density for ECG signal');
```

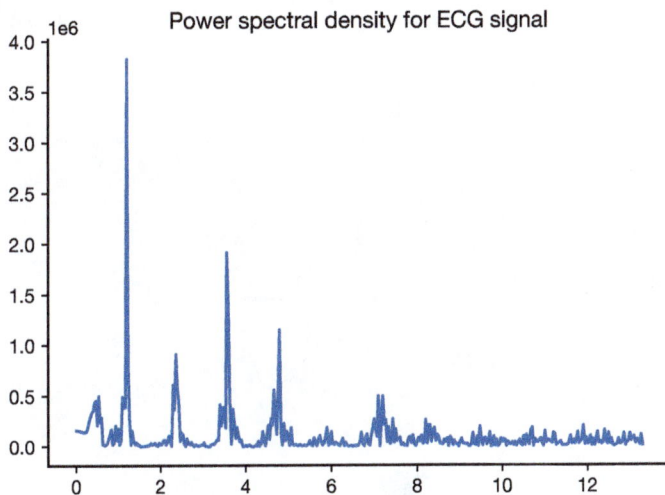

In most real applications, we would estimate the PSD by averaging over multiple time windows of the signal – see the documentation for `scipy.signal.ShortTimeFFT()` for details. However, we will see that the simple technique shown here works well for this example.

Let's zoom in on the portion of the ECG signal that contains the frequency with the highest power:

```
plt.plot(f[:80], np.abs(ECG[:80])**2)
plt.title('Power spectral density for ECG signal');
```

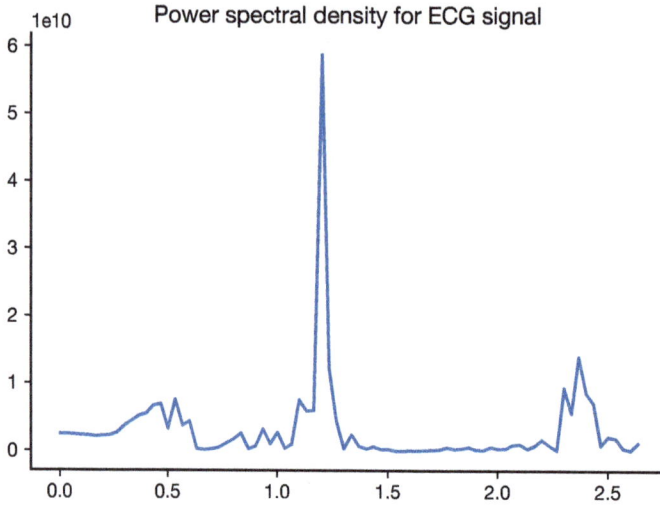

We can get the exact index of the peak using the `argmax()` method:

```
peak_pos = (ECG[:400]**2).argmax()
peak_pos
```

36

and the corresponding frequency is:

```
f[peak_pos]
```

1.2000203032472179

This frequency is in Hertz. This frequency corresponds to the heart rate during the ECG, which is approximately 1.2 beats/second, or 72 beats per minute.

For a periodic non-sinusoidal signal, the DFT will contain harmonics at multiples of the base frequency. Thus, in the power spectral density, we see peaks not only at 1.2 Hz but also at 2.4 Hz, 3.6 Hz, and 4.8 Hz. Those other peaks are related to the same 1.2 Hz electrical signal that is driving the beating of the heart at 72 beats/minute. Although we will not pursue further analysis of the ECG signal in this book, some research articles do look at the power spectral density at the lower frequencies below the heartbeat.

The DFT introduced in this section is one example of a universal basis that is commonly used because it is a powerful and simple technique for transforming a time-domain signal into a frequency-domain representation.

Terminology review and self-assessment questions

Interactive flashcards to review the terminology introduced in this section and self-assessment questions are available at la4ds.net/6-2, which can also be accessed using this QR code:

6.3 Set-Specific Bases: The Gram-Schmidt Algorithm

In the previous section, we showed one example of universal bases that consist of complex sinusoids: the rows of DFT matrices. The advantage of a universal basis is that it can represent every vector in some vector space $\mathbb{R}^n$. The disadvantage is that the cardinality of a universal basis in $\mathbb{R}^n$ is always n.

In this section, we consider finding a set-specific basis and show that such a basis can have a much smaller cardinality than a universal basis. Let's motivate the general approach using an example.

Example 6.9: Finding a Set-Specific Basis for Four 8-Vectors

Consider the vectors $\mathcal{S}_a = \{s_0, s_1, s_2, s_3\}$, where

$$
\begin{aligned}
s_0 &= \begin{bmatrix} 1 & 1 & 1 & 1 & -1 & -1 & -1 & -1 \end{bmatrix}, \\
s_1 &= \begin{bmatrix} 0 & 2 & 2 & 0 & 0 & -2 & -2 & 0 \end{bmatrix}, \\
s_2 &= \begin{bmatrix} 1 & -1 & -1 & 1 & -1 & 1 & 1 & -1 \end{bmatrix}, \text{ and} \\
s_3 &= \begin{bmatrix} 2 & 0 & 0 & 2 & -2 & 0 & 0 & -2 \end{bmatrix}.
\end{aligned}
$$

These vectors are from $\mathbb{R}^8$, so a universal basis for span$(\mathcal{S}_a)$ would consist of 8 vectors. We know that $\mathcal{S}_a$ is a spanning set for span$(\mathcal{S}_a)$, and this implies that $\dim\big(\text{span}(\mathcal{S}_a)\big) \leq |\mathcal{S}_a| = 4$. Let's see if we can find a basis for span$(\mathcal{S}_a)$.

Let $\mathbf{f}_i$ be the ith basis vector. We iterate through the signals one-by-one. We will end up with different bases depending on the order in which we iterate over the signals, but for this example, we will iterate over them in numerical order.

Signal s_0

We start with s_0 and create the basis vector $\mathbf{f}_0$ by normalizing it:

```
import numpy as np
from numpy.linalg import norm


s0 = np.array([1,  1,  1,  1, -1, -1, -1, -1])
f0= s0 / norm(s0)
print(f0)
```

```
[ 0.35355339  0.35355339  0.35355339  0.35355339 -0.35355339 -0.35355339
 -0.35355339 -0.35355339]
```

Note that $\|\mathbf{s}_0\|^2 = 8$, so it is easier to see $\mathbf{f}_0$ mathematically as

$$\mathbf{f}_0 = \frac{1}{\sqrt{8}} \begin{bmatrix} 1 & 1 & 1 & 1 & -1 & -1 & -1 & -1 \end{bmatrix}.$$

Then the projection of $\mathbf{s}_0$ onto $\mathbf{f}_0$ is

$$\begin{aligned}
\langle \mathbf{s}_0, \mathbf{f}_0 \rangle &= \langle \mathbf{s}_0, \frac{\mathbf{s}_0}{\|\mathbf{s}_0\|} \rangle \\
&= \frac{1}{\|\mathbf{s}_0\|} \langle \mathbf{s}_0, \mathbf{s}_0 \rangle \\
&= \frac{\|\mathbf{s}_0\|^2}{\|\mathbf{s}_0\|} \\
&= \|\mathbf{s}_0\| \\
&= \sqrt{8} \approx 2.828.
\end{aligned}$$

Let's check numerically:

```
s0 @ f0
```

```
2.82842712474619
```

Signal $\mathbf{s}_1$

To find a second basis vector, we start by finding the scalar projection of $\mathbf{s}_1$ onto the basis vector $\mathbf{f}_0$:

```
s1 = np.array( [ 0,  2,  2,  0,  0, -2, -2,  0] )

s1 @ f0
```

```
2.82842712474619
```

Then the part of $\mathbf{s}_1$ that cannot be represented by $\mathbf{f}_0$ can be found by subtracting the vector projection of $\mathbf{s}_1$ onto $\mathbf{f}_0$ from the vector $\mathbf{s}_1$:

```
e1 = a1 - (a1 @ f0) * f0
print(e1)
```

```
[-1.  1.  1. -1.  1. -1. -1.  1.]
```

which has norm

```
norm(e1)
```

2.8284271247461903

Since the error signal has a nonzero norm, we need another basis function that is orthogonal to $\mathbf{f}_0$. Fortunately, the error signal is always orthogonal to the previous basis functions. Let's check for this example:

```
np.round(e1 @ f0, 10)
```

0.0

Then to create our second basis vector, we can simply normalize the error vector $\mathbf{e}_1$:

```
f1 = e1 / norm(e1)
print(f1)
```

```
[-0.35355339  0.35355339  0.35355339 -0.35355339  0.35355339 -0.35355339
 -0.35355339  0.35355339]
```

As with $\mathbf{f}_0$ it is easier to write this mathematically as

$$\mathbf{f}_1 = \frac{1}{\sqrt{8}} \begin{bmatrix} -1 & 1 & 1 & -1 & 1 & -1 & -1 & 1 \end{bmatrix}.$$

The projection of $\mathbf{s}_1$ onto $\mathbf{f}_1$ is

```
s1 @ f1
```

2.8284271247461907

which is the norm of $\mathbf{e}_1$.

Signal $\mathbf{s}_2$

Note that it turns out that we could also write

$$\mathbf{f}_1 = -\frac{1}{\sqrt{8}} \mathbf{s}_2.$$

Another way to interpret this is that $\mathbf{s}_2$ is in the (opposite) direction of $\mathbf{f}_1$. Thus, $\mathbf{s}_2$ should be orthogonal to $\mathbf{f}_0$ and so the scalar product of $\mathbf{s}_2$ with $\mathbf{f}_0$ should be zero:

```
s2 = np.array( [1, -1, -1,  1, -1,  1,  1, -1] )

s2 @ f0
```

0.0

We can see that we can write $s_2 = -\sqrt{8}f_1$, which matches the result from the numerical projection,

```
print(-(f1 * np.sqrt(8)))
print( s2.astype(float))
```

```
[ 1. -1. -1.  1. -1.  1.  1. -1.]
[ 1. -1. -1.  1. -1.  1.  1. -1.]
```

There is no remaining error, so we do not need any new basis function.

Signal s_3

Last, we try to represent s_3. The scalar projection of s_3 onto f_0 is

```
s3 =np.array( [ 2,  0,  0,  2, -2,  0,  0, -2] )

s3 @ f0
```

2.82842712474619

and the scalar projection of s_3 onto f_1 is

```
s3 @ f1
```

-2.8284271247461894

The error signal can be calculated by subtracting the sum of the vector projections from s_3,

```
e3 = s3 - ( (s3 @ f0)*f0 + (s3 @ f1)*f1 )
np.round(e3, 10)
```

```
array([ 0.,   0.,   0.,   0.,  -0.,  -0.,  -0.,  -0.])
```

Since the error signal is the zeros vector, we do not need to create a new basis vector.

Signal Set Dimensionality

Two basis vectors are sufficient to represent all the signals in $\mathcal{S}_a$. In the context of communications, we say that the dimension of this signal set is 2. (More generally, the dimension of the vector space $\text{span}(\mathcal{S}_a)$ is two.) In comparison, a universal basis would require 8 basis vectors.

6.3.1 Signal Space Representation

Given a signal set $\mathcal{S}$, every signal in $\mathcal{S}$ can be expressed as a linear combination of the vectors in our orthonormal basis. If there are N basis functions $\mathbf{f}_0, \mathbf{f}_1, \ldots, \mathbf{f}_{N-1}$, then we can write

$$\mathbf{s}_i = \sum_{k=0}^{N-1} s_{ik} \mathbf{f}_k.$$

Here, each coefficient of the linear combination is given by the scalar projection of $\mathbf{s}_i$ onto the basis $\mathbf{f}_k$, which is $s_{ik} = \mathbf{s}_i \cdot \mathbf{f}_k$. Then given a particular basis, each signal can be represented by a vector of the coefficients that multiply the basis function, $\bar{\mathbf{s}}_i = [s_{i,0}, s_{i,1}, \ldots, s_{i,N-1}]$. The vector $\bar{\mathbf{s}}_i$ is called the *signal-space representation* of $\mathbf{s}_i$:

DEFINITION

signal-space representation

If $\mathcal{S}$ is a set of vectors over $\mathbb{R}^n$ with orthonormal basis $\mathcal{F} = \{\mathbf{f}_0, \mathbf{f}_1, \ldots, \mathbf{f}_{N-1}\}$, where

$$\mathbf{s}_i = \sum_{k=0}^{N-1} s_{ik} \mathbf{f}_k,$$

then the vector $\bar{\mathbf{s}}_i = [s_{i,0}, s_{i,1}, \ldots, s_{i,N-1}]$ is the *signal-space representation* of $\mathbf{s}_i$.

We will use $\overline{\mathcal{S}}$ to denote the set of signal-space representations.

Note:

The signal-space representations depend on the orthonormal basis used to represent the signals, and the basis is not unique. However, we will see soon that the properties of the signal-space representation do not depend on the choice of basis.

To understand this better, let's apply it to our example 4-ary signal set:

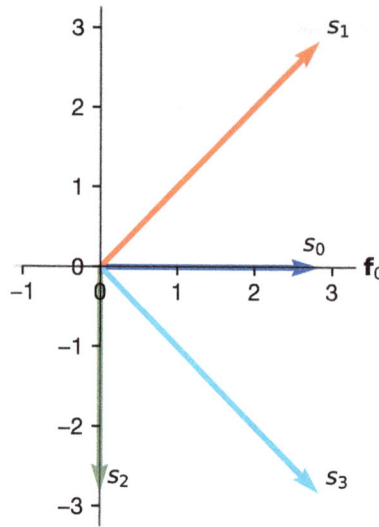

Fig. 6.5: Signal-space representation for set of four vectors from $\mathbb{R}^8$.

Example 6.10: Signal Space Representation for Example 4-ary Signal Set

Consider again the 4 signal vectors from Example 6.9 along with the orthonormal basis we found, $\mathcal{F} = \{\mathbf{f}_0, \mathbf{f}_1\}$. Then the signal-space representation for $\mathbf{s}_i$ is

$$\bar{\mathbf{s}}_i = [\mathbf{s}_i \cdot \mathbf{f}_0, \ \mathbf{s}_i \cdot \mathbf{f}_1].$$

Even though the original vectors were of length 8, the signal-space representations are of length 2, because all of the signals can be represented as linear combinations of two basis functions. Since the new representation occupies only two dimensions, we can now plot these vectors. The result is shown in Fig. 6.5.

6.3.2 Properties of Signal-Space Representations

An important feature of signal-space representations is that they preserve the most important aspects of the original signal set. Consider a set of signals $\mathcal{S}$ and corresponding signal-space representations $\bar{\mathcal{S}}$.

Then the signal-space representations have the following properties:

1. **Inner-product preserving:**

$$\text{For all } i, k, \quad \langle \mathbf{s}_i, \mathbf{s}_k \rangle = \langle \bar{\mathbf{s}}_i, \bar{\mathbf{s}}_k \rangle.$$

2. **Norm preserving:**

$$\|\mathbf{s}_i\| = \|\bar{\mathbf{s}}_i\|.$$

3. Distance preserving:

$$\text{For all } i, k, \quad \|\mathbf{s}_i - \mathbf{s}_k\| = \|\bar{\mathbf{s}}_i - \bar{\mathbf{s}}_k\|.$$

Properties 2 and 3 follow directly from Property 1. These properties ensure that our signal-space representation is meaningful. For instance, in our plot of the vectors using the signal-space representation, the lengths of the vectors are the norms of the original vectors, and we can measure the distance between vectors as the length of a vector connecting the heads of those vectors.

Let's check properties 2 and 3 for our example vectors:

Example 6.11: Checking Norm- and Distance-Preserving Property of Signal-Space Representation

Consider the signals and their signal-space representations from Example 6.9. The following Python code compares the norms of the signals to the norms of their corresponding signal-space representations:

```python
print('Property 2: Norm Preserving\n')
print(' i | Norm of original vector | Norm of signal-space rep')
print('-'*55)

for i in range(4):
  print(f'{i:^3}|{norm(A[i]):^25.1f}|{norm(Areps[i]):^25.1f}')
```

```
Property 2: Norm Preserving

 i | Norm of original vector | Norm of signal-space rep
--------------------------------------------------------
0 |            2.8           |            2.8
1 |            4.0           |            4.0
2 |            2.8           |            2.8
3 |            4.0           |            4.0
```

The following code compares the distances between all the original signal vectors, $d_{ij} = \|\mathbf{s}_i - \mathbf{s}_j\|$, to the distances between their signal-space representations, $\bar{d}_{ij} = \|\bar{\mathbf{s}}_i - \bar{\mathbf{s}}_j\|$:

```python
print('Property 3:\n')
print(' i, k | Dist for original vectors | Dist for signal-space reps')
print('-'*62)
for i in range(4):
  for k in range(0,4):
    print(f'{i:>2}, {k:<2}|{norm(A[i] - A[k]):^27.1f}|'
          +f'{norm(Areps[i] - Areps[k]):^27.1f}')
  print()
```

```
Property 3:

i, k | Dist for original vectors | Dist for signal-space reps
-------------------------------------------------------------
0, 0 |           0.0            |            0.0
0, 1 |           2.8            |            2.8
0, 2 |           4.0            |            4.0
0, 3 |           2.8            |            2.8

1, 0 |           2.8            |            2.8
1, 1 |           0.0            |            0.0
1, 2 |           6.3            |            6.3
1, 3 |           5.7            |            5.7

2, 0 |           4.0            |            4.0
2, 1 |           6.3            |            6.3
2, 2 |           0.0            |            0.0
2, 3 |           2.8            |            2.8

3, 0 |           2.8            |            2.8
3, 1 |           5.7            |            5.7
3, 2 |           2.8            |            2.8
3, 3 |           0.0            |            0.0
```

6.3.3 Gram-Schmidt Process

The procedure we conducted above for finding an orthonormal basis will work for **any** set of vectors. It is called the Gram-Schmidt Process, and the general algorithm is shown below:

Given indexed vectors $s_0, s_1, \ldots, s_{K-1}$.

1. Let $i = 0$. Let $\mathcal{F} = ()$ be the ordered collection of basis vectors (initialized to empty).

2. For $j = 0, \ldots, |F| - 1$: calculate the scalar projection of s_i onto each of the basis vectors: $s_{ij} = s_i \cdot f_j$.

3. Calculate the error vector e_i, which is the part of s_i that is orthogonal to all the basis vectors up to this point: $e_i = s_i - (s_{i0}f_0 + s_{i1}f_1 + \ldots)$.

4. If $\|e_i\| = 0$, then s_i can be completely represented in terms of the basis vectors in $\mathcal{F}$. Increment i (i.e., $i = i + 1$) and go to step 2.

5. Else normalize the error vector to create a new basis vector, $f_{|\mathcal{F}|} = e_i / \|e_i\|$ and go to step 2.

In practice, step 4 needs to be modified to check if $\|e_i\| < \epsilon$ because limits of floating-point arithmetic often result in values that should be zero returning some small value instead.

6.3.4 Dimensionality and Linear Independence

The dimension of a set of signals is the same as the maximum number of linearly independent vectors in the set. We previously showed that we can find the maximum number of linearly independent vectors by stacking the vectors into the columns (or rows) of a matrix and using `np.linalg.matrix_rank()`. Let's check this with our example set of four 8-vectors:

Example 6.12: Dimensionality of Set of 8-Vectors Using Matrix Rank

Let's create a matrix **S** whose columns are the vectors s_0, s_1, s_2, s_3 from Example 6.9. As we've seen before, NumPy treats vectors like the row of a matrix, so stacking them horizontally with `np.hstack()` will result in one long vector. Instead, we stack them vertically using `np.vstack()` and then transpose the result to end up with the vectors in columns:

```
S = np.vstack((s0, s1, s2, s3)).T
S
```

```
array([[ 1,  0,  1,  2],
       [ 1,  2, -1,  0],
       [ 1,  2, -1,  0],
       [ 1,  0,  1,  2],
       [-1,  0, -1, -2],
       [-1, -2,  1,  0],
       [-1, -2,  1,  0],
       [-1,  0, -1, -2]])
```

Then the matrix rank is:

```
np.linalg.matrix_rank(S)
```

```
2
```

Since the maximum rank of an 8×4 matrix is $\min(8, 4) = 4$, this matrix is rank deficient.

If a matrix has full rank, then we can find a basis for the columns of the matrix using the NumPy command `np.linalg.qr()`, which gives the QR decomposition of the matrix. If the input is a matrix **M**, then the output is $\mathbf{M} = \mathbf{Q} \cdot \mathbf{R}$, where **Q** has orthogonal columns that are a basis for **M**.

However, if the matrix is rank deficient, the QR decomposition may not yield the minimum number of orthogonal basis functions. In general, it is better to use the SciPy function `scipy.linalg.orth()` to find a basis for the columns. The resulting basis vectors are in the columns of the returned matrix.

Example 6.13: Finding a Basis Using `scipy.linalg.orth()`

We can find a basis for our example by passing the S matrix as the sole argument of `scipy.linalg.orth()`:

```
import scipy.linalg
Q = scipy.linalg.orth(S)
print(Q)
```

```
[[-0.35355339 -0.35355339]
 [ 0.35355339 -0.35355339]
 [ 0.35355339 -0.35355339]
 [-0.35355339 -0.35355339]
 [ 0.35355339  0.35355339]
 [-0.35355339  0.35355339]
 [-0.35355339  0.35355339]
 [ 0.35355339  0.35355339]]
```

To make this easier to compare with the basis we found using the Gram-Schmidt algorithm, let's take the transpose and multiply by $\sqrt{8}$. Below that matrix, we will print the basis functions that we previous found, also scaled up by $\sqrt{8}$.

```
print(np.sqrt(8) * Q.T)
```

```
[[-1.  1.  1. -1.  1. -1. -1.  1.]
 [-1. -1. -1. -1.  1.  1.  1.  1.]]
```

```
print(np.sqrt(8) * f0)
print(np.sqrt(8) * f1)
```

```
[ 1.  1.  1.  1. -1. -1. -1. -1.]
[-1.  1.  1. -1.  1. -1. -1.  1.]
```

The first column of $\mathbf{Q}$ is the same as $\mathbf{f_1}$, and the second column of $\mathbf{Q}$ is the same as $-\mathbf{f_0}$. Bases are not unique, but we can see that they are very similar in this case.

In Fig. 6.6, I show the representation we found using the Gram-Schmidt procedure and the representation using the basis from scipy.linalg.orth(). The second representation is equivalent to a 90° clockwise rotation of the first. Although the representations are different, the norms and distances are the same using either representation.

Gram-Schmidt basis   scipy.linalg.orth() basis

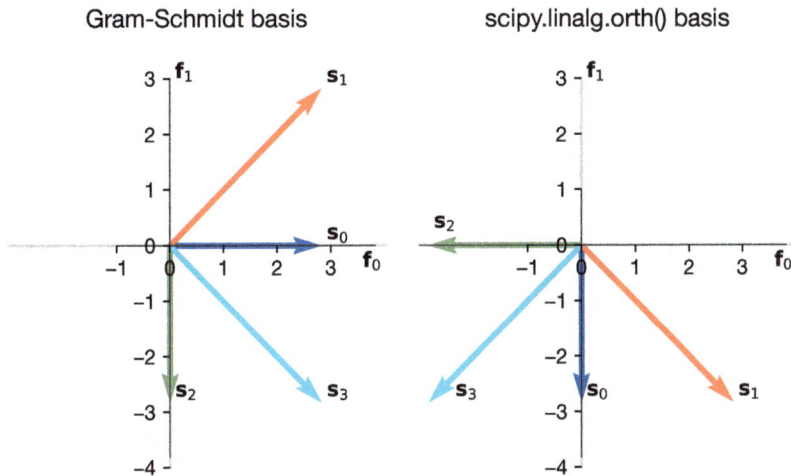

Fig. 6.6: Two signal-space representations of the four signals from Example 6.9, based on two different bases.

Example 6.14: Detecting Communication Signals

Modern wireless communication systems use digital communications, in which one or more bits of information are used to select a communication waveform to send. Quadrature phase-shift keying (QPSK) is a common signaling scheme that is used in both WiFi wireless local area networks and in cellular communication systems, such as LTE and 5G. QPSK conveys two bits in each signaling interval. In its simplest form, the waveforms for QPSK look like:

$$s_0(t) = A\cos\left(\omega t + \frac{\pi}{4}\right)p(t) \qquad s_1(t) = A\cos\left(\omega t - \frac{\pi}{4}\right)p(t)$$

$$s_2(t) = A\cos\left(\omega t + \frac{3\pi}{4}\right)p(t) \qquad s_3(t) = A\cos\left(\omega t - \frac{3\pi}{4}\right)p(t).$$

Here, A and ω control the amplitude and frequency, respectively, of the signal; $p(t)$ is a signal that limits the duration of the signal to one bit time.

These signals are shown in Fig. 6.7 for $A = 1$, $\omega = 2\pi$, and

$$p(t) = \begin{cases} 1, & 0 \le t \le 1 \\ 0, & \text{otherwise.} \end{cases}$$

A digital receiver samples the received signal at multiple samples per symbol. In the absence of noise, the sampled signals are shown below for a sampling rate of 10 samples/symbol. The following code creates an array for which column i contains the samples of signal i:

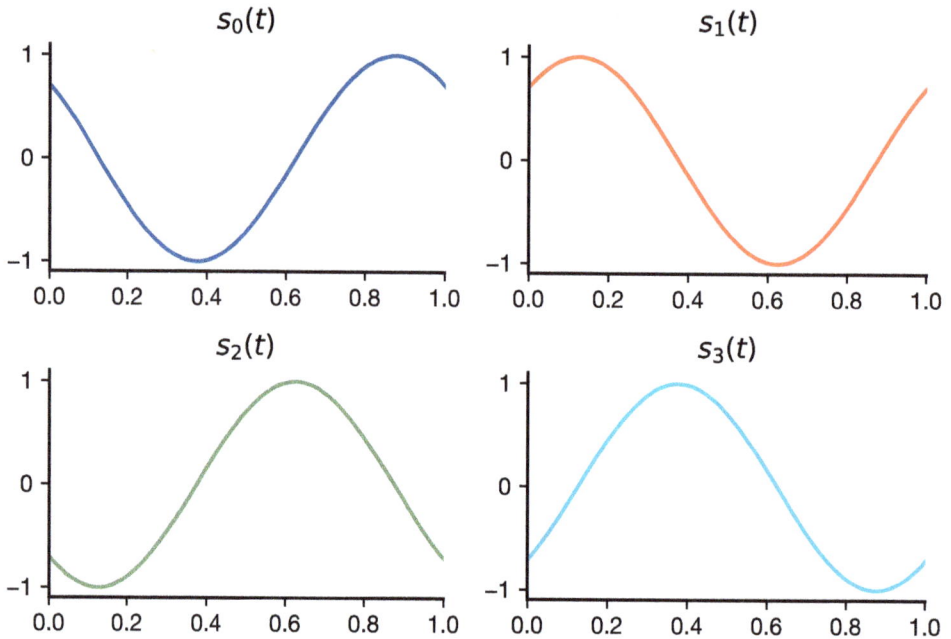

Fig. 6.7: Example of four QPSK digital modulation signals.

```
phases = [np.pi/4, -np.pi/4, 3*np.pi/4, -3*np.pi/4]

signals = np.zeros((10, 4))
t2 = np.linspace(0+1/20, 1+1/20, 10)
for signum in range(4):
  signals[:,signum] = np.cos(2*np.pi*t2 + phases[signum])
```

These vector signals are shown with each component's value plotted as a dot versus its index in Fig. 6.8. We can find a basis for these signals using `scipy.linalg.orth()`:

```
sig_basis = scipy.linalg.orth(signals)
print(sig_basis)
```

```
[[ 4.26401433e-01  7.45228750e-16]
 [ 3.26642448e-01 -3.03012985e-01]
 [ 7.40438317e-02 -4.64242827e-01]
 [-2.13200716e-01 -4.08248290e-01]
 [-4.00686280e-01 -1.61229842e-01]
 [-4.00686280e-01  1.61229842e-01]
 [-2.13200716e-01  4.08248290e-01]
 [ 7.40438317e-02  4.64242827e-01]
```

(continues on next page)

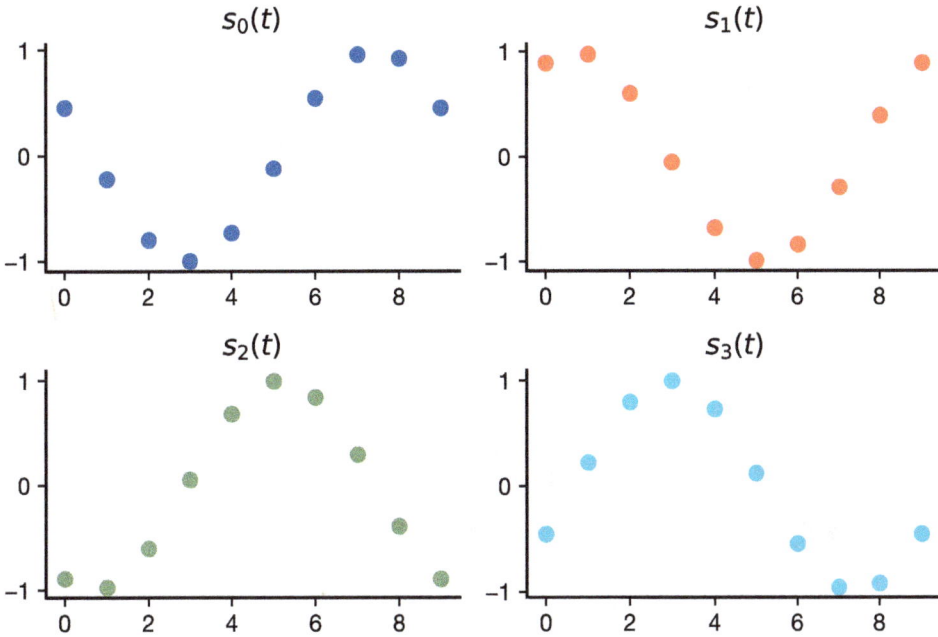

Fig. 6.8: Vectors representing four different QPSK symbols.

(continued from previous page)

```
[ 3.26642448e-01   3.03012985e-01]
[ 4.26401433e-01   8.39150995e-16]]
```

The signals have dimension 2, and the basis functions found are essentially sampled cosine and sine waves, as shown in Fig. 6.9.

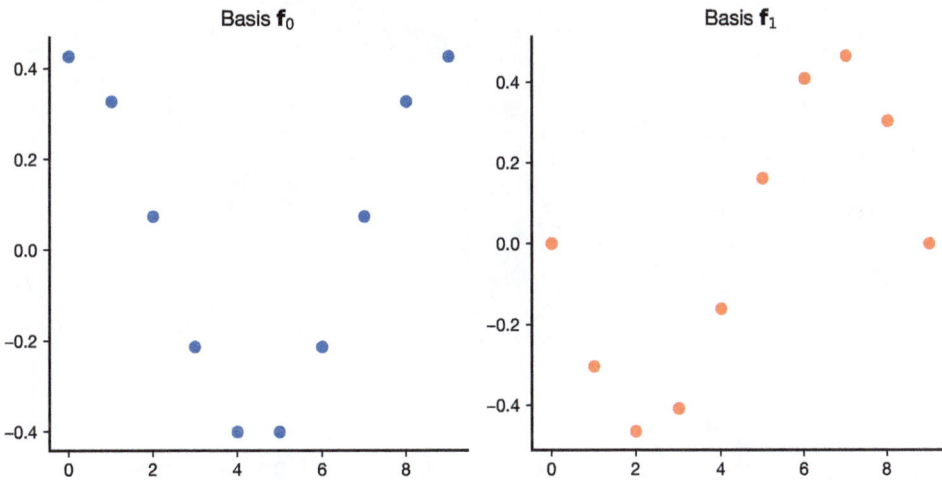

Fig. 6.9: Basis vectors for QPSK signal vectors.

Signal-space representations are commonly used for communication signals. For one- and two-dimensional signal sets, we often illustrate the signals by showing their signal-space representations as points on a line or plane. These are called *signal constellations*. A signal constellation for this signal set is shown in Fig. 6.10. In most books on digital communications, the basis is chosen such that the constellation points are at $\pm 45°, \pm 135°$.

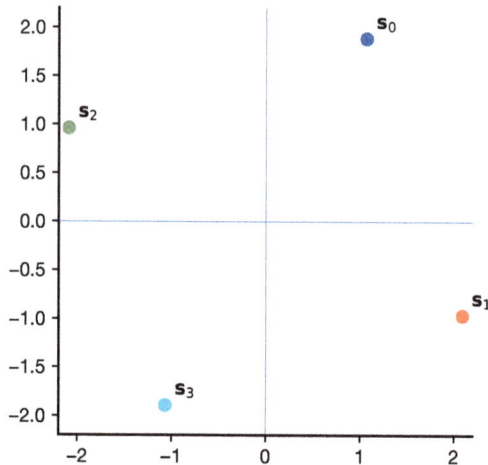

Fig. 6.10: Signal constellation for QPSK.

One of the big advantages of using this signal-space representation is that it can be used to make decisions about a noisy signal. The thermal noise in receivers results in the samples of the received signals being corrupted by noise from a Normal (i.e., Gaussian) distribution. Let's create and plot an example of receiving the signal $s_2(t)$ in the presence of thermal noise:

```python
import scipy.stats as stats
np.random.seed(7)

N = stats.norm(0, 1)
noise = N.rvs(10)
r = signals[:,2] + noise
plt.scatter(range(10), r)
plt.title('Received samples of signal $s_2(t)$ in presence of thermal noise');
```

Received samples of signal $s_2(t)$ in presence of thermal noise

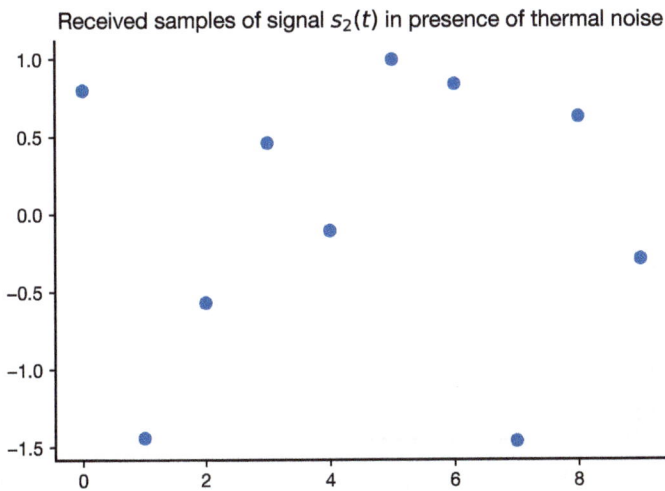

The purpose of a receiver is to take the received noisy samples and decide which signal was sent. From looking at the plot of the example noisy signal, it seems unclear how this decision should be made. From a visual comparison of the samples of the received signal and the possible transmitted signals, it seems difficult to know which signal was sent. To enable us to make a decision, let's project the received signal onto our basis:

```
r0 = r @ sig_basis[:,0]
r1 = r @ sig_basis[:,1]
print(f'The signal space representation of r is [{r0 : .3g},{r1 : .3g})']
```

```
The signal space representation of r is [-0.831, 0.544]
```

This creates a received vector **r**. Fig. 6.11 shows the signal constellation along with the signal-space projection of the received signal (the × mark annotated with **r**). The plot shows that **r** is closest to the signal-space representation s_2, and so that is the best decision.

The following code calculates the part of the received signal that lies within the signal space and the part of the received signal that lies outside of the signal space:

```
fig, axs = plt.subplots(1, 2, figsize=(8,4) )

axs[0].scatter(range(10), (r0*sig_basis[:,0] + r1*sig_basis[:,1]))
axs[0].set_title('Parts of received signal within the signal space');

axs[1].scatter(range(10), r - (r0*sig_basis[:,0] + r1*sig_basis[:,1]))
```

(continues on next page)

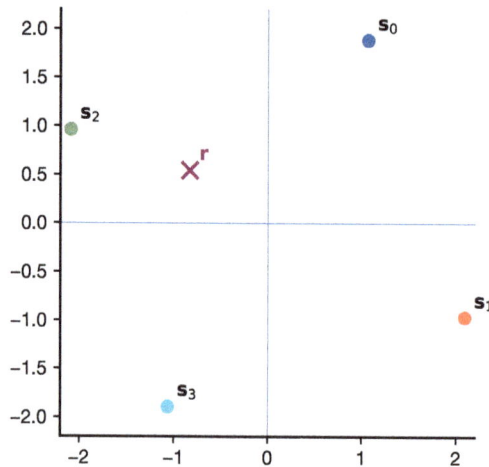

Fig. 6.11: QPSK signal constellation (dots) with example noisy received signal's signal-space representation ($\times$).

(continued from previous page)

```
axs[1].set_title('Parts of received signal outside of signal space');

plt.tight_layout()
```

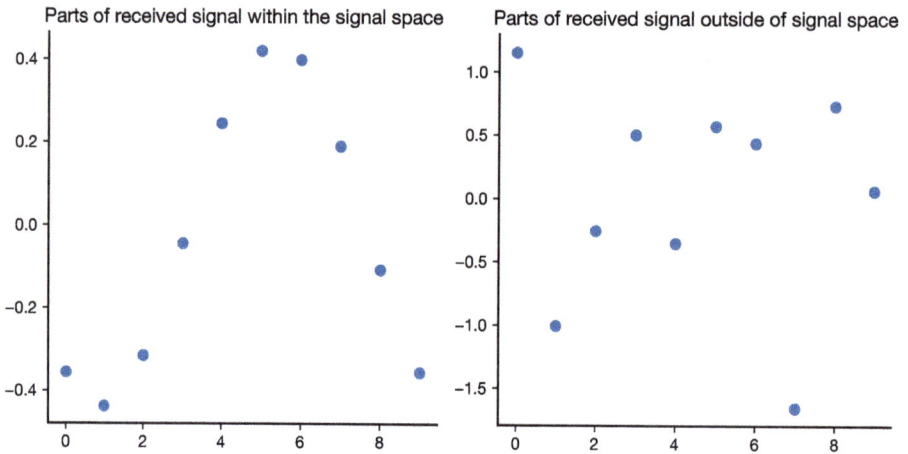

Note that the part of the signal that lies within the signal space looks quite similar to s_2, which corresponds to the optimal decision. Moreover, the part of the signal that lies outside the signal space corresponds to noise, and this noise is effectively removed from the signal by projecting the received signal into the signal space. Because all of the parts of the received signal that contain information about the transmitted signal are preserved in the signal-space representation, an optimal decision can be made.

In the next section, we look at a simple example of how an alternative basis can be used in classification when we do not have a model for the data.

Terminology review and self-assessment questions

Interactive flashcards to review the terminology introduced in this section and self-assessment questions are available at la4ds.net/6-3, which can also be accessed using this QR code:

6.4 Alternative Bases via Eigendecomposition

Consider again the problem of feature extraction, which was originally introduced in Section 3.2.5. In that section, we showed how we can use matrix-vector multiplication to rotate a two-dimensional data set to extract a feature that can be used for classifying data as belonging to one of two classes. However, the method introduced there was *ad hoc* (literally, "for this") – it does not provide a general method to solve other problems, and it is not necessarily optimal in any sense. In this section, we develop a more systematic approach to address this issue by leveraging our knowledge of bases and eigendecomposition.

From our previous work on eigenvalues and eigenvectors, we know that if an $n \times n$ matrix $\mathbf{M}$ has n linearly independent eigenvectors, then the matrix can be written as $\mathbf{M} = \mathbf{U}\mathbf{\Lambda}\mathbf{U}^{-1}$. Then we can also *diagonalize* the matrix $\mathbf{M}$ as

$$\begin{aligned} \mathbf{U}^{-1}\mathbf{M}\mathbf{U} &= \mathbf{U}^{-1}\left(\mathbf{U}\mathbf{\Lambda}\mathbf{U}^{-1}\right)\mathbf{U} \\ &= \left(\mathbf{U}^{-1}\mathbf{U}\right)\mathbf{\Lambda}\left(\mathbf{U}^{-1}\mathbf{U}\right) \\ &= \mathbf{I}\mathbf{\Lambda}\mathbf{I} \\ &= \mathbf{\Lambda}. \end{aligned}$$

This might be a useful property if it somehow allows us to use the eigenvalues to extract the information down to a simpler form. One issue is that matrices do not even have eigenvalues and eigenvectors unless they are square, which is generally not the case for most data sets. For instance, the Iris data set considered in Section 3.2.5 consists of 150 data points, each of which has four features, so we can represent it by a 150×4 or 4×150 matrix.

Rather than try to decompose the data directly, let's consider a matrix that measures the variation and dependence among the different features of the data and see whether we can use that information to find a new basis for representing the data. The first step in measuring "variation" of the data is to find some point the data is varying around. We will use the average (mean) of each feature as the point around which we measure the variation. If we collect the means into a vector, it is called the *mean vector*:

> **DEFINITION**
>
> **mean vector (data),**
> **sample mean**
>> Consider an $m \times n$ matrix of data D, where each column represents a data point and each row represents a feature. Then the *mean vector* is the average of the columns,
>>
>> $$\bar{\mathbf{d}} = \frac{1}{n}\sum_{k=0}^{n-1}\mathbf{d}_k.$$

Example 6.15: Mean Vector for Iris Data

Let's load the Iris data from `scikit-learn` and compute the mean vector. Since the default in `scikit-learn` is that the data points are in rows of the `data` matrix, we will transpose that data and store it into a variable called `DIris`. Then we will use `np.mean()` to compute the average. By default `np.mean()` computes the average of all of the data. To average by feature, we use the keyword argument `axis=1` to indicate to average over the different data points, which are in the columns (axis 1):

```
from sklearn import datasets

iris = datasets.load_iris()
DIris = iris.data.T

np.mean(DIris, axis=1)
```

```
array([5.84333333, 3.05733333, 3.758, 1.19933333])
```

The ith entry in the vector is the average value of the data for feature i. It is important to note that different features have different averages. This may be inherent in the type of feature (for instance, in this example the features correspond to measurements of different parts of the Iris plant) or it may also be caused by other factors, such as choice of unit. For instance, all of these Iris measurements are in units of cm, but if one of them was measured in mm, then the mean for that feature would be 10 times higher than if it had been measured in cm. In addition, measurements like this may be subject to offsets based on how the measurements were conducted. For instance, features 0 and 3 are sepal length and petal length; different data sets might vary on whether the length was measured to the stem or to the end of the sepal/petal after removal. Later, we will consider ways to remove some of these effects from our data.

Once we have the mean vector, we can calculate the variations of the features away from their means. Let's start with the simplest of these, which is called the variance. It is simply the average squared distance of the data for a feature from the mean for that feature:

DEFINITION

variance (data),
sample variance

Consider an $m \times n$ matrix of data D, where each column represents a data point and each row represents a feature. Let the mean vector be denoted $\overline{\mathbf{d}}$, and let the average of the ith feature be denoted $\overline{d}_i$. Then the (unbiased) *variance* (or *sample variance*) of feature i is

$$s_i^2 = \frac{1}{n-1} \sum_{k=0}^{n-1} \left(d_{i,k} - \overline{d}_i \right)^2 .$$

Note that the division by $n-1$ in this definition is different than the usual average of dividing by n. Some variance definitions use division by n. But dividing by $n-1$ gives a nice property called unbiasedness, and so we will use division by $n-1$ in this book.

Example 6.16: Variance of Iris Data

Consider again the Iris data stored in the NumPy array `DIris`. We can get the variance for all of the features using the NumPy function `np.var()`, but we need to pass several keyword arguments to get the desired result. As with `np.mean()`, we need to specify to only average over the columns using the keyword argument `axis=1`. To use the divisor of $n-1$, we need to specify the keyword argument `ddof=1`, where ddof stands for "delta degrees of freedom", specifying how much smaller than n the divisor should be. Thus, the variances are:

```
np.var(DIris, axis=1, ddof=1)
```

```
array([0.68569351, 0.18997942, 3.11627785, 0.58100626])
```

In Section 3.2.5, we introduced the concept of feature extraction. A simpler approach is called *feature selection*:

DEFINITION

feature selection

In *feature selection*, only a subset of the features present in the data are used or preserved.

Feature selection and feature extraction are two approaches used for *dimensionality reduction*:

> **DEFINITION**
>
> **dimensionality reduction**
>
> The process of going from a high-dimensional data set to a lower-dimensional representation of that data set, usually with the goal of preserving as much important information as possible from the original data set.

The variance can give us insight into which features should be preserved if we use feature selection. If the variance of a feature is low, then, in general, that feature can be well approximated by its mean. On the other hand, features with large variance take on much more diverse values. To preserve the greatest amount of information about the data, we should generally preserve those features with high variance if using feature selection.

Variance does not tell the whole story because a data set my have features that have high variance but where the features are highly dependent on each other. For instance, we may be able to use one feature to accurately predict the values of the other features. This motivates using a feature-extraction approach that can extract the most important information from the input features. To do so, we need information not only on the variance of the features but also measures of the dependence among the features. A common statistic that is used for this is a generalization of the variance called *covariance*:

> **DEFINITION**
>
> **covariance (data),**
> **sample covariance**
>
> Consider an $m \times n$ matrix of data $\mathbf{D}$, where each column represents a data point and each row represents a feature. Let the mean vector be denoted $\overline{\mathbf{d}}$, and let the average of the data for the ith feature be denoted $\overline{d}_i$. Then the (unbiased) *covariance* (or *sample covariance*) between feature i and feature j is
>
> $$\mathrm{Cov}\left(\mathbf{d}_i, \mathbf{d}_j\right) = \frac{1}{n-1} \sum_{k=0}^{n-1} \left(d_{i,k} - \overline{d}_i\right)\left(d_{j,k} - \overline{d}_j\right).$$

In general, larger covariances (relative to the variances) indicate that the features are more related to each other in the sense that one variable can be well predicted using a linear predictor based on the other variable. If the covariance between two features is zero, we say that those features are *uncorrelated*. For our purposes, if two features are uncorrelated, it means that we cannot use a linear function of one feature to estimate the other feature – that will not give us any additional information about the feature being estimated. Additional interpretation is outside the scope of this book, but readers are referred to *Foundations of Data Science with Python*, also by John M. Shea.

All of the covariances among the features can be calculated efficiently using matrix operations, but for our purposes, we will use `np.cov()`. No keyword arguments are necessary because the default divisor for this function is already $n-1$. The results is a *covariance matrix* where entry i, j is the covariance between features i and j. The i, ith entry of the covariance matrix is the variance of feature i. Let's test this on the Iris data:

Example 6.17: Covariance of Iris Data Features

The covariance matrix for the Iris data set is:

```
K = np.cov(DIris)
K
```

```
array([[ 0.68569351, -0.042434  ,  1.27431544,  0.51627069],
       [-0.042434  ,  0.18997942, -0.32965638, -0.12163937],
       [ 1.27431544, -0.32965638,  3.11627785,  1.2956094 ],
       [ 0.51627069, -0.12163937,  1.2956094 ,  0.58100626]])
```

Let's start with some observations that are specific to this example:

- The diagonal of the covariance matrix contains the variances we previously found with `np.var()`.

- Some pairs of features have much larger covariances than others. For instance, the covariance between features 0 and 1 is approximately -0.04, whereas the covariance between features 2 and 3 is approximately 1.30.

We can see some properties generalize to any covariance matrix:

Properties of Covariance Matrices

- The diagonal of a covariance matrix corresponds to the variances.

- A covariance matrix is square and symmetric.

- Unlike variance, covariances can be positive or negative. Positive covariance indicate that the features "tend to move in the same direction", whereas negative features "tend to move in opposite directions".

In the following, we consider only a special case of covariance matrices, where the determinant is strictly positive. In this case, an $n \times n$ covariance matrix will have n positive eigenvalues, and the modal matrix $\mathbf{U}$ will be an orthogonal matrix. Recall that for an orthogonal matrix, $\mathbf{U}^T\mathbf{U} = \mathbf{U}$, so $\mathbf{U}^{-1} = \mathbf{U}^T$. Thus, a covariance matrix $\mathbf{K}$ can be factored as $\mathbf{K} = \mathbf{U}\mathbf{\Lambda}\mathbf{U}^T$. In addition, for data matrix $\mathbf{D}$ with covariance matrix $\mathbf{K}$, it can be shown that the covariance matrix of $\tilde{\mathbf{D}} = \mathbf{U}^T\mathbf{D}$ is

$$\tilde{\mathbf{K}} = \mathbf{U}^T\mathbf{K}\mathbf{U}$$
$$= \mathbf{U}^T\left(\mathbf{U}\mathbf{\Lambda}\mathbf{U}^T\right)\mathbf{U}$$
$$= \mathbf{\Lambda}.$$

Thus, the resulting data become uncorrelated, and the variances of the transformed data are equal to the eigenvalues of $\mathbf{K}$. We say that we *decorrelated* the data through this linear transform.

Let's test this using the Iris data:

Example 6.18: Decorrelating the Iris Data

Let's start by finding the eigendecomposition of the covariance matrix for the Iris data and printing out the eigenvalues. Because the covariance matrix is symmetric, it is best to use the `la.eigh()` function to find the eigendecomposition:

```
lam, U = la.eigh(K)
lam
```

```
array([0.02383509, 0.0782095 , 0.24267075, 4.22824171])
```

Now let's transform the data by left-multiplying by $\mathbf{U}^T$ and calculate the covariance matrix of the transformed data:

```
Dt = U.T @ DIris

np.round(np.cov(Dt), 3)
```

```
array([[ 0.024, -0.   , -0.   , -0.   ],
       [-0.   ,  0.078, -0.   , -0.   ],
       [-0.   , -0.   ,  0.243,  0.   ],
       [-0.   , -0.   ,  0.   ,  4.228]])
```

We can see that the off-diagonal elements of the transformed data are all zero, and the diagonal elements equal the eigenvalues of the covariance matrix. Note that the largest variance after the transformation, 4.228, is bigger than the largest variance in the original data, 3.116. Similarly, the smallest variance after the transformation, 0.024, is smaller than the smallest variance in the original data, 0.190.

This approach of left-multiplying the data by the transpose of the modal matrix of the covariance matrix can be considered to be a form of the discrete Karhunen-Loève Transform (KLT). In fact, the KLT results in some even more useful properties:

- The KLT achieves the projection with the largest variance possible for projection onto a unit-norm vector. The resulting variance is the largest eigenvalue of the covariance matrix.

- If we consider maximizing the remaining variances by projecting onto unit-norm vectors that are orthogonal to the ones previously found, the resulting variances will be the eigenvalues in decreasing order.

- The above properties result in the minimum possible variance achievable by projection onto a unit-norm vector being achieved by the KLT and equal to the minimum eigenvalue.

These properties are very useful for dimensionality reduction because we can preserve features that have as large a variance as possible over all features that can be created with orthogonal linear projections of the data, and the resulting features will be uncorrelated. Let's apply this concept to visualize the Iris data:

Example 6.19: 2-D Visualization of the 4-D Iris Data

The Iris data set has four features (dimensions), so there is no way to directly visualize the data using a scatter plot. However, if we transform the data as shown in Example 6.18, we can then drop the two features with very low variances and make a scatter plot of the remaining features. The result is shown in Fig. 6.12, where I have used different markers to distinguish the different classes present in the data. Feature 3 has the largest variance and is clearly the most useful in distinguishing between the different classes.

Note that the KLT has projected the data in such a way that we can easily distinguish between the 'setosa' and 'versicolor' classes using KLT feature 3 – the KLT has created a feature that achieves the same goal as the *ad hoc* approach we showed in Section 3.2.5, but using an approach that can be generalized to arbitrary data sets. Feature 3 can also be used to perform most of the distinction between the 'versicolor' and 'viriginica' classes, but more sophisticated classifiers that use both features 2 and 3 will perform better.

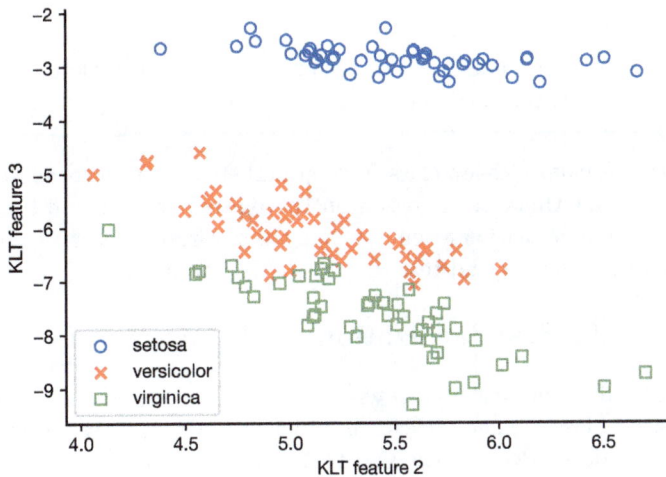

Fig. 6.12: Scatter plot of data using two features with largest variance at output of KLT.

The combined approach of decorrelation and dropping low-variance data is called *principal components analysis (PCA)*:

> ## DEFINITION
>
> **principal components analysis (PCA)**
>
> A dimensionality reduction technique that consists of
>
> 1. calculating the sample covariance matrix, $\mathbf{K}$,
> 2. performing eigendecomposition on $\mathbf{K}$ to get the eigenvalue vector $\boldsymbol{\lambda}$ and the modal matrix $\mathbf{U}$,
> 3. projecting the data matrix $\mathbf{D}$ onto the columns of the modal matrix as $\mathbf{U}^T\mathbf{D}$, resulting in uncorrelated data with covariance matrix $\boldsymbol{\Lambda}$, and
> 4. dropping some number of output features with the lowest variance.

One problem with directly applying KLT to the data covariance matrix is that it is sensitive to the absolute magnitudes of the features. Features with higher variance are more likely to play an important role in the output features. However, the variances of an input feature can be easily changed just by expressing it in different units (for instance, if the sepal width in the Iris data set were recorded in mm or μm instead of cm, the variance would be much larger). To overcome this, we often *standardize* the data before applying PCA (or other machine-learning algorithms):

> ## DEFINITION
>
> **standardization**
>
> The process by which numerical data is transformed such that each feature has mean zero and variance one.

We will use the `StandardScaler` class from `scikit-learn`'s `preprocessing` class to standardize data. Note that this class is based on the biased covariance estimator, so we will pass the keyword argument `ddof=0` when checking the covariance matrix of the output. Let's illustrate the use of this object and test this on the Iris data.

Example 6.20: Standardization of the Iris Data

To standardize the Iris data, we must first instantiate an object with the `StandardScaler` class. Then we can use the object's `fit_transform()` method to standardize the data. Because the `scikit-learn` methods expect data to be in rows, we will transpose the Iris data in the argument of that method and again on the method's output:

```
from sklearn.preprocessing import StandardScaler
scaler = StandardScaler()
Ds = scaler.fit_transform(DIris.T).T


np.mean(Ds, axis=1), np.cov(Ds, ddof=0)
```

```
(array([-1.69031455e-15, -1.84297022e-15, -1.69864123e-15, -1.40924309e-15]),
 array([[ 1.        , -0.11756978,  0.87175378,  0.81794113],
        [-0.11756978,  1.        , -0.4284401 , -0.36612593],
        [ 0.87175378, -0.4284401 ,  1.        ,  0.96286543],
        [ 0.81794113, -0.36612593,  0.96286543,  1.        ]]))
```

The transformed data has zero mean vector and unit variances.

We can apply the KLT to the standardized data in the same way that we did when the data is not standardized. However, the usual way to perform PCA on standardized data is using a different transform called the *singular value decomposition (SVD)*:

DEFINITION

singular value decomposition (SVD)

Every real $m \times n$ (i.e., not just square) matrix $\mathbf{M}$ can be factored as

$$\mathbf{M} = \mathbf{U}\mathbf{\Sigma}\mathbf{V}^T, \tag{6.2}$$

where $\mathbf{U}$ and $\mathbf{V}$ are orthogonal matrices, and $\mathbf{\Sigma}$ is a $m \times n$ matrix with the only non-zero entries on the main diagonal. The values on the diagonal of $\mathbf{\Sigma}$ are called singular values, and they are similar to eigenvalues of square matrices.

For standardized data, we can get the same effect as KLT by applying the transformation $\mathbf{U}_S^T\mathbf{D}$ to the data, where $\mathbf{U}_S$ is the left singular-vector matrix from the SVD.

Example 6.21: Two Approaches to KLT of Standardized Iris Data

Let's decorrelate the standardized Iris data in two different ways. First, we use eigendecomposition of the covariance matrix:

```
Ks = np.cov(Ds)
lams, Us = la.eigh(Ks)
lams
```

```
array([0.02085386, 0.14774182, 0.9201649 , 2.93808505])
```

```
np.round(np.cov(Us.T @ Ds), 3)
```

```
array([[ 0.021, -0.   ,  0.   ,  0.   ],
       [-0.   ,  0.148,  0.   , -0.   ],
       [ 0.   ,  0.   ,  0.92 ,  0.   ],
       [ 0.   , -0.   ,  0.   ,  2.938]])
```

We see that the data is decorrelated, and the variances are equal to the eigenvalues of
the covariance matrix. We could again use this in PCA and plot the highest variance
components, but the result looks similar to that when the data is not standardized.
Thus, this is left as an exercise for the reader.

Now let's apply the SVD to the data and then use the $\mathbf{U}_S$ matrix to transform
the standardized data:

```
Usvd, Ssvd, Vsvd = la.svd(Ds)


np.round(np.cov(Usvd.T @ Ds), 3)
```

```
array([[ 2.938,  0.   , -0.   ,  0.   ],
       [ 0.   ,  0.92 ,  0.   ,  0.   ],
       [-0.   ,  0.   ,  0.148, -0.   ],
       [ 0.   ,  0.   , -0.   ,  0.021]])
```

The data is again decorrelated, and the variances are equal to the eigenvalues of the
covariance matrix. However, the ordering of the output features is changed from
that using eigendecomposition of the covariance matrix.

When using NumPy's `la.svd()`, the singular values are sorted in descending order. This
makes it easy to apply PCA by preserving the first features in the data. (For `la.eigh()`,
the eigenvalues are in increasing order, and the eigenvalues for `la.eig()` are not necessarily
ordered.)

As simple as KLT/PCA is, we can also perform it directly using the `PCA` class from
scikit-learn's `decomposition` module.

Example 6.22: Performing KLT/PCA Using `scikit-learn`

The code below shows how to apply PCA by creating an object of the `PCA` class
and then transforming the data using the `fit_transform()` method. Keep in mind
that if each column corresponds to a data point, then the data matrix needs to be
transposed at the input and output of this method.

```
from sklearn.decomposition import PCA


pca = PCA()
Ds_pca = pca.fit_transform(Ds.T).T
np.round(np.cov(Ds_pca), 3)
```

```
array([[ 2.938,  0.   ,  0.   ,  0.   ],
       [ 0.   ,  0.92 , -0.   , -0.   ],
       [ 0.   , -0.   ,  0.148, -0.   ],
       [ 0.   , -0.   , -0.   ,  0.021]])
```

The result matches that from using the SVD, and is equivalent to the result from using the modal matrix of the covariance matrix.

Terminology review and self-assessment questions

Interactive flashcards to review the terminology introduced in this section and self-assessment questions are available at la4ds.net/6-4, which can also be accessed using this QR code:

6.5 Chapter Summary

In this chapter, I introduced the concepts of universal and set-specific bases for collections of vectors and showed the power of projecting data onto different bases. First, I showed how to create a sinusoidal basis for n-vectors using the DFT matrix, and we applied this to determine the frequency of a person's heartbeat captured from an electrocardiogram (ECG). Next, I showed how to find a set-specific basis using the Gram-Schmidt procedure, and we investigated the application of this to determining the transmitted signal from a noisy received signal in a digital communication system. Finally, I showed how we can transform a data set to make the features uncorrelated and to identify the maximum-variance features that can be created using linear combinations of the input features. This is often used in principal components analysis (PCA), which is a dimensionality reduction technique. Multiple approaches to achieve this were shown using eigendecomposition, singular-value decomposition, and the `PCA` class from `scikit-learn`. We applied PCA to visualize the four-dimensional Iris data set by plotting the two features with highest variance after decorrelation.

Access a list of key take-aways for this chapter, along with interactive flashcards and quizzes at la4ds.net/6-5, which can also be accessed using this QR code:

Index

For Product Safety Concerns and Information please contact our EU
representative GPSR@taylorandfrancis.com
Taylor & Francis Verlag GmbH, Kaufingerstraße 24, 80331 München, Germany

www.ingramcontent.com/pod-product-compliance
Lightning Source LLC
Chambersburg PA
CBHW061936190326
41458CB00009B/2752